The Global Assault on Teaching, Teachers, and their Unions

Mary Compton is the past President of the National Union of Teachers, the largest teacher union in the UK and Europe. She is the author of many articles in education and teacher union journals, including *Rethinking Schools* and *New Politics*. She is also a playwright, whose latest play on climate change is part of a project for the British Department for Food and Rural Affairs.

Lois Weiner is Professor of Elementary and Secondary Education at New Jersey City University. She is an internationally known authority on urban teaching and the author of *Preparing Teachers for Urban Schools: Lessons from Thirty Years of School Reform* (Teachers College Press, 1993); *Urban Teaching: The Essentials* (Teachers College Press, 2006 second edition). She is the author of numerous articles in scholarly journals and popular publications about urban education and teacher unions.

The Global Assault on Teaching, Teachers, and their Unions
Stories for Resistance

Mary Compton and Lois Weiner, Editors

palgrave
macmillan

First published in 2008 by
PALGRAVE MACMILLAN™
175 Fifth Avenue, New York, N.Y. 10010 and
Houndmills, Basingstoke, Hampshire, England RG21 6XS.
Companies and representatives throughout the world.

PALGRAVE MACMILLAN is the global academic imprint of the Palgrave
Macmillan division of St. Martin's Press, LLC and of Palgrave Macmillan Ltd.
Macmillan® is a registered trademark in the United States, United Kingdom
and other countries. Palgrave is a registered trademark in the European
Union and other countries.

ISBN-13: 978-0-230-60631-9(hardcover)
ISBN-10: 0-230-60631-8(hardcover)
ISBN-13: 978-0-230-60630-2(paperback)
ISBN-10: 0-230-60630-X (paperback)

Library of Congress Cataloging-in-Publication Data

The global assault on teaching, teachers, and their unions : stories for resistance
/ Mary Compton and Lois Weiner, editors.
  p. cm.
Includes bibliographical references and index.
ISBN 0-230-60630-X (alk. paper)—ISBN 0-230-60631-8 (alk. paper) 1.
Privatization in education—Cross-cultural studies. 2. Teachers' unions—
Cross-cultural studies. 3. Education and globalization—Cross-cultural studies. 4.
Neoliberalism—Cross-cultural studies. I. Compton, Mary F. II. Weiner, Lois.

LB2806.36.G58 2008
331.88'113711—dc22   2007037225

A catalogue record of the book is available from the British Library.

Design by Scribe Inc.

First edition: April 2008

10 9 8 7 6 5 4 3 2 1

Printed in the United States of America.

Lois and Mary dedicate this book to teachers all over the world who are defending public education, and, with it, human emancipation and hope.

Contents

Vignettes

Acknowledgments

The depth and breadth of this volume has been made possible by the labor of our contributors, and we thank them.

From Lois: I thank my husband and comrade, Michael Seitz, for his moral and political support and ready ear and my daughter, Petra, for reminding me that students can provide us with compelling insights about schools. My participation in the AFT local in Hayward, California, especially work with Lew Hedgecock, its president for many years, demonstrated the progressive possibilities of teacher unionism. Many years spent active in the New York City local of the AFT and more recently in the higher education local of AFT at New Jersey City University have informed my thinking about what's needed to democratize and reinvigorate unions. This book reflects the ideas and idealism of many wise and committed union members with whom I've worked over these years, too numerous to name. They have inspired and instructed me, as has the international assemblage of scholars in the "Teachers' Work/Teacher Unions" Special Interest Group of the American Educational Research Association. A grant from the office of the Vice President of Academic Affairs at New Jersey City University provided release time from teaching to complete research for this volume. All of these sources of assistance have made my participation in this project possible, but the ideas I present are my sole responsibility.

From Mary: I would like to thank my husband Hugh Pope and daughters Blanche, and Faith for their support while I was carrying out this project at a difficult time. Thanks also to my daughters Clarrie and Helen for all their help with editing and for their patience. I would like to thank the National Union of Teachers and its members and activists for their comradeship in waging the fight for state education, which has never ceased to inspire me, and its leaders and staff for helping me in my time of office as president of the union, without which my eyes would never have been opened to the magnitude and coherence of the neoliberal assault. However in coediting and writing in this book I am representing my own views and writing in a personal capacity. Lastly, I am profoundly grateful to all the colleagues with whom I have taught over the years, whose hard work on behalf of children has given me the inspiration to fight with them for our rights and our futures. In particular I thank my job-share partner Clare Victor, who has always inspired my teaching and who supported me unstintingly through my period of office.

PART I

Neoliberalism, Teachers, and Teaching
Understanding the Assault

CHAPTER 1

The Global Assault on Teachers, Teaching, and Teacher Unions

Mary Compton and Lois Weiner

Teachers in every part of the world are in the forefront of the struggle to ensure that children receive an education—whether in U.S. cities, the mountains of Chavez's Venezuela, in civil war-torn Nepal, in Europe's towns and countryside, or in the refugee camps of Sudan. In prosperous nations, identified by global justice activists as the global north, teachers' wages, their voice in policy, and the quality of their working conditions have been reduced. In the south, societies that lack power and wealth in the new global economy, teachers are working for a pittance, sometimes unpaid, too often poorly trained or with no training at all. In many parts of Africa they teach children in classes of over 100, sometimes under trees or squatting in churches. Textbooks are nonexistent, libraries and computers a distant dream, and basic materials in short supply. When wars occur, declared or not, schools are taken over and the children evicted, as has been done in Congo and by the Israeli army in the Palestinian West Bank. In Sri Lanka and the Indian Ocean islands, schools were wiped out by the tsunami, and while luxury resorts are being built, school construction lags. In impoverished city neighborhoods in the "developed" world, from Paris and London to Berlin and New York, teachers are struggling against reduced school funding, racism, and social and economic dislocation to try to ensure that all children receive quality schooling.

As a classroom teacher elected President of the British National Union of Teachers (NUT), Mary had the unique opportunity to visit teachers' conferences all around the world. The almost overwhelming social and political crises that

---

Vignette 1.1

"The Book of Knowledge Investing in the Growing Education and Training Industry," Merrill Lynch Report, *April 9, 1999*:

> A new mindset is necessary, one that views families as customers, schools as "retail outlets" where educational services are received, and the school board as a customer service department that hears and addresses parental concerns. As a near monopoly, schools escape the strongest incentives to respond to their customers—the discipline of the market.

---

so many teachers face were debated at all of these events. Activists at the 2002 South African Democratic Teachers Union (SADTU) Congress in Durban debated the merits of various funeral plans for members and their families who would be killed by HIV and AIDS, the lack of water and electricity in their communities, school fees for the poor, and delays in payment of salaries. SADTU members were often working in appalling conditions, all too often sick and dying and teaching children in similar straits. In addition to dealing with these social crises, which are exacerbated by policies of world financial institutions, teachers must simultaneously contend with the deterioration in public education, itself an outcome of reforms promoted and imposed by those very same institutions. Though the titles and acronyms of policies differ from one country to another, the basics of the assault are the same: undercut the publicly supported, publicly controlled system of education, teachers' professionalism, and teacher unions as organizations. For example, much of the SADTU conference was taken up with issues that are also facing teachers in the UK—performance-related pay, wasteful and bureaucratic inspection processes, and increasing privatization. The conference of the Australian Educators Union (AEU) devoted a large part of its time to issues of teacher quality and how this could be monitored. The Canadian Teachers' Federation conference Mary attended wrestled with the issues of national testing, monitoring, league tables (published comparisons of school test results), and inspections. Teacher unionists in Germany debated a whole raft of government measures touted as promoting quality control, a panacea for the problems in the German education system, which, according to official reports, is failing a large number of children of both the poor and middle classes. The very nature of education is being contested: the Fourth World Congress of the international organization of teacher unions, Education International (EI), held in Brazil, explored the theme "Education: Public Service or Commodity?"

Enshrined in the Universal Declaration of Human Rights, a covenant of the United Nations, is an affirmation that education is a basic right of all children The millennium development goal for education set in Dakar, Senegal, in the year 2000 is universal primary education by 2015. Despite the apparent desire of the world's leaders to make it possible for all children to be educated, many communities, schools, children, and teachers confront reforms that sabotage the explicit commitment to basic education for all.

Over the last couple of decades a new global consensus about reshaping economies and schools has emerged among the politicians and the powerful of the world. Whereas in the past governments—preferably democratically elected—have assumed the responsibility to ensure that all children are educated, schools and universities are now regarded as a potential market. In these educational markets, entrepreneurs set up schools and determine what is taught and how it is taught in order to make a profit. The assumption that schooling is a "public good" is under the most severe attack it has ever endured. As the contributions to this volume demonstrate, teacher trade unionists are grappling with the increasing privatization of education services, the introduction of business "quality control" measures into education, and the requirement that education produce the kind of minimally trained and flexible workforce that corporations require to maximize their profits. Among scholars and global justice activists, these reforms being made to the economy and education are often called "neoliberal." (Susan Robertson explains this terminology in her essay following this chapter.) Though the term "neoliberalism" may be new to many readers, the policies are not. They are experienced almost universally by teachers, children, and parents.

Many governments under the sway of neoliberal ideas have relinquished pursuit of the policies needed to provide a basic education to all, chief among these financing education adequately so that all teachers can teach and learners learn in conditions fit for that purpose. Clearly the millennium development goals for education will not be met. While rich northern nations spend billions of dollars prosecuting wars and have bottomless resources for the exploiting of new gas and oil reserves, the most precious reserves of all—the world's children—stand at the back of the line. Nor is there an opportunity to develop education systems so that they can fulfill their true purpose—to enable people to live a full and creative life, or as the Universal Declaration of Human Rights puts it, to ensure that education is directed "toward the full development of the human personality."

There is an old saying that "a lie gets halfway round the world before the truth gets on its running shoes." The lie making the running in schooling is that private corporations and entrepreneurs are much more able to make education work for the poor than teachers, communities, and their elected representatives in government. And when one listens to politicians and reads in the media about the benefits of bringing the private market and business methodologies into education, one can often feel like teachers have hardly begun to tie the laces on their running shoes. The voices for privatization and neoliberalism have virtually the whole of the world's media at their disposal to speed them on their way. Rebutting the "private good, public bad" propaganda is complicated by neoliberalism's hijacking of ideals and terms borrowed from those who have spent their lives campaigning for education for all and opportunities for the poor and oppressed. Hearing news reports and politicians' statements of lofty goals, one might think there is nothing closer to the hearts of the international financiers, accountants, and politicians than the needs of the poor. It is only when you look at the actual effects of the policies of world financial institutions like the International Monetary Fund (IMF) and World Bank on "developing" countries and their education systems that you realize that nothing could be further from the truth. As many researchers, including those writing in this book, document, the World Bank's structural adjustment programs have destroyed perfectly adequate education systems in countries like Zambia and are threatening to do the same in many others. As readers will see in the excerpts from neoliberal reports, websites, and corporate financial bulletins with titles like "Why school fees are good for the poor," when it serves their purpose, neoliberal gurus are quite willing to ditch the rhetoric of social justice and equality and lay bare the true face of their education policy.

Teachers are in a war being fought over the future of education, and though at times it might seem as though we are losing the war without firing a shot, we have a potentially powerful weapon in our hands—our solidarity and organization into powerful teachers unions. EI, which brings the teaching unions of the world together, has over 29 million members. Although researchers and activists engage in a lively global discourse about the need to wrest education back from the private entrepreneurs and corporations who want to turn it into a cash cow and a source of flexible labor, the debate contains very little discussion—from nongovernmental organizations that do advocacy work (NGOs), from academics, and even from unions themselves—of the role that teacher unions can and must play in reversing these policies. Our shared commitment to illuminate the neglected role of teacher trade unions, to analyze their potential power to win an alternative type of education, underlies our work of bringing together the essays in this book.

Public service unions, and in particular education unions, do have power. And yet it often feels as though the tremendous potential force that is contained in those 29 million teachers organized into trade unions is not being used. The economic system that dominates the world, capitalism, has become global—its strategy and propaganda have a global reach and logic. Yet we in the trade union movement, despite some traditions of international solidarity, are a long way from achieving an equivalent global coherence. And even on a national level trade union leaders are often too ready to accept the rhetoric of politicians as the reality and adapt accordingly, instead of standing up boldly and opposing them—if necessary through job action.

Ironically, the potential power of teachers and our unions to derail neoliberal reforms like privatization is often more apparent to our opponents than it is to teachers and union leadership. In a draft report for the World Bank, "Making Services Work for Poor People," economist Ritva Reinikka argues that teaching unions impede equitable development of services for poor people by diverting scarce resources toward teachers' pay. Reinikka cites teachers' failure to "perform their duties in a reasonable fashion" as "probably the biggest constraint to making services work for poor people." Conveniently omitted in this and other World Bank analyses is the fact that teachers' pay is appallingly low and sometimes nonexistent in many developing countries.

In the face of this kind of rhetoric and in the context of the global assault by private capital on state education services, how are the unions responding, and how should they work to defend services for the poor? Is there any indication that teacher unions and their leaderships have the will to face up to this situation and reverse it? These are the issues our book takes up, through contributions by teachers and researchers throughout the world. The book is informed by the firsthand knowledge Mary and Lois have had in their careers as classroom teachers and union activists, as well as by their different roles outside the classroom. Since leaving her job as a New York City public school teacher, Lois has been a college professor and researcher who educates urban

---

Vignette 1.2

From the World Bank report, "A Chance to Learn: Knowledge and Finance for Education in Sub-Saharan Africa," http://www.worldbank .org/afr/hd/wps/chance_learn.pdf (accessed August 12, 2007):

> In particular, recent progress in primary education in Francophone countries resulted from reduced teacher costs, especially through the recruitment of contractual teachers, generally at about 50% the salary of civil service teachers. . . . All IFC education investments must provide a satisfactory financial return. The definition of "satisfactory" is dependent on IFC-wide strategic investment directions.

teachers and studies teachers' work and unions. Mary has been the president of England's largest teacher union and has worked with teacher unionists in much of the globe. We have seen evidence, in research and practice, that a more assertive and democratic teacher union movement is possible and can turn back the neoliberal agenda.

In her travels, Mary has seen many straws in the wind. In Durban, the end of the SADTU conference signaled the beginning of a mass strike and demonstration for a maximum class size of fifty in the KwaZulu Natal province. The leaders of the union marched from the conference podium to the steps of the town hall to demand more money for education. The Egitem Sen teachers' conference in Ankara, Turkey, was also followed by a mass demonstration—this time against the threatened closing down of the union by a government that was refusing to recognize the rights of its Kurdish minority to be taught in its own language. The conference of a French teacher union, SNES, was interrupted by a strike and mass demonstrations all over France for the restoration of teachers' pension rights. In Wales, the combined effect of a boycott of standardized compulsory tests for young children and league tables of schools, along with the concerted lobbying of parents and teacher unions, brought about their abandonment. And at EI, a motion condemning the invasion of Iraq was passed overwhelmingly despite opposition from the U.S. teacher unions.

As a new classroom teacher, Lois helped organize a local of the American Federation of Teachers (AFT) in California. Later, when she began to teach in New York City, she became active in efforts to democratize the local in New York, the AFT's largest. Teaching at the university, Lois sees in prospective teachers an idealism about making a difference in children's lives, like the commitment that animated teachers in the United States to create teacher unions in the 1960s and 70s. Now a researcher, Lois participates in the Special Interest Group on teachers' work and teacher unions of the American Educational Research Association. Several authors in this collection belong to this international network of scholars, which has raised the visibility of teacher unions in educational reform and begun to clarify how teacher unions can be revitalized.

In compiling this book, we interspersed scholarly analyses with personal, firsthand accounts of the impact that neoliberal policies are having on teachers around the world. The book explains what neoliberal education policy is and why it is so important that teachers in general and union activists in particular understand its rationale and implications. Teachers, union activists, union officials, and researchers from around the globe write about a range of pressing problems that have no simple solutions. We have undertaken this ambitious project in the hope that it will be a useful tool for activists in education who, as we do, want teacher unions to use their power to fight for and bring to fruition the ideal of a quality education for every child in the globe.

Mary Compton, Wales
Lois Weiner, New York City
August 2007

CHAPTER 2

"Remaking the World"
Neoliberalism and the Transformation of Education and Teachers' Labor

Susan L. Robertson

In the introduction to his short book, a brief history of neoliberalism, David Harvey (2005, 1) observes "Future historians may well look upon the years 1978–80 as a revolutionary turning point in the world's social and economic history." Four significant events acted like epicenters in the unfolding of the transformation of the postwar order: in 1978, Deng Xiaoping took the first steps toward liberalizing the Chinese economy; in 1979, Paul Volcker took command of the U.S. Federal Reserve and changed monetary policy while in that same year Margaret Thatcher took on the power of the unions and pledged to end inflationary stagnation. Harvey writes, "In 1980, Ronald Reagan was elected President of the United States and, armed with Volcker's policies, set about implementing a set of reforms that were aimed at curbing union power, deregulating industry, and creating more liberal conditions for finance to operate on the national and the global stage. From these several epicenters, revolutionary impulses seemingly spread and reverberated to remake the world around us in a totally different image" (Harvey 2005, 1).

Three decades later, few disagree that the globalization of what is for neoliberalism a utopia has become fact. Its promoters have remade the world, including the world of education. Out with the collective and welfare; in with the individual and freedom. This tectonic shift, "like continental drift," (Bourdieu 1998, 1) has transformed how we talk about education, teachers and learners, unions, parents' groups, and professional associations. In short, it has altered the conditions for knowledge production and the circumstances under which we might demand a socially just education system, along with the spaces and sites for claims making around education. With education yoked more closely to national and regional economies, schools and universities are now universally mandated to (efficiently and effectively) create the new breed of entrepreneurs and innovators, the value-driven minds who will spearhead the battle for global markets and consumers and a bigger share of the spoils. Education, once mostly untrammeled virgin territory, is also being initiated into the world of property rights, markets, trade, and rating agencies (Molnar 2006; Hentschke 2006).

How might we explain this transformation of education in the post-World War II settlement? A core argument of this chapter is that the mobilization of neoliberal ideas for reorganizing societies and social relations, including the key institutions involved in social reproduction, is a class project of capitalism with three key aims: (1) the redistribution of wealth upward to the ruling elites through new structures of governance, (2) the transformation of education systems so that the production of workers for the economy is the primary mandate, and (3) the breaking down of education as a public sector monopoly, opening it up to strategic investment by for-profit firms. To be realized, all three aims must break down the institutionalized interests of teachers, teacher unions, and fractions of civil society who have supported the idea of education as a public good and public sector, an intrinsic element of the social contract between the state and civil society.

There is, however, also considerable evidence that the rolling out of neoliberal policies and programs across the globe has been highly uneven. The creation of this new world system has experienced tectonic jolts because of its own internal contradictions (cf. Fine 2001), because the histories and institutions of places cannot be airbrushed out of existence (Fourcade-Gourinchas and Babb 2002; Madrid 2003; Harvey 2006; Ong 2006), and because in myriad ways, neoliberalism has been visibly and vigorously contested (cf. Waterman 2001). The presence of contradictions, cultures, and contestation does not, of course, enable us to wind back the clock, as if the events that make up history can be made not to matter. Nor will I argue that the neoliberal program has delivered entirely negative outcomes. As always, life is more complicated than that. However, I will argue that neoliberalism has transformed, albeit in both predictable and unpredictable ways, *how* we think and *what we do* as teachers and learners, and it is therefore important we make these things evident to ourselves.

In doing so, I hope we might see the *limits* of the closer relation between education and capitalism, and the centrality of class, labor, and exploitation to our explanations of what is going on. This understanding will enable us to confront, overcome, and transcend those limits. In this introductory chapter I look back to the beginning of this epochal change to clarify what, at the time, was not so clear: the nature of the project that was underway and, more particularly, what the struggle portends.

Neoliberalism's Origins

Neoliberalism did not arrive unannounced from the shadows, though it felt this way. At the time of the events that announced its arrival, I was working with a group of Head teachers employed by the Department of Education, Western Australia, on a very progressive initiative that had devolved some power, authority, and accountability to the school and its community. What in retrospect were the first neoliberal reforms were disarmingly called School Development. However, they had a very different feel to them from the progressive policies we were attempting to work with and write about. As I studied the new policies, part of the New Public Management initiatives, I detected that they had more to do with economics than real efficiency, effectiveness, or public and managerial accountability.

Though they seemed to arrive unannounced, neoliberalism's fundamental ideas are not particularly new. Rather, neoliberalism's pedigree can be traced to liberalism, a utopian project promoted by philosophers such as Locke and Hobbes, who were committed to the ideals of personal freedom and possessive individualism. Liberalism also stood opposed to collectivism. The core ideas of liberalism are outlined by Macpherson (1962): individuals have freedom from dependence on others, the individual is the proprietor of his own person and capacities, human societies consist of a series of market relations, and political society is constructed to protect an individual's property and goods. In other words, supreme value is given to individual autonomy, agency, and property. Neoliberalism can be understood as a variant of liberalism. However, neoliberalism differs from liberalism in one important way; its commitment to a different type of economics, termed "neoclassical," that recognizes that some state intervention is necessary to ensure that Adam Smith's hidden hand of the market can function. This means that in contrast to liberalism, neoliberalism demands that freedom of the market, the right to free trade, the right to choose, and protection of private property be assured by the state.

Neoliberalism should be contrasted with the "marriage between economic liberalism and social democracy" during the golden years of the welfare state (Hobsbawm 1994, 270). As Hobsbawm noted, post–World War II capitalism was unquestionably a system reformed out of all recognition. With the welfare state flourishing, little ground was given to the proponents of neoliberalism. Keynesian socioeconomic policies were widely deployed, especially in the thirty countries belonging to the Organization for Economic Co-operation and Development (OECD). However, by the 1960s things had begun to change. The balanced compromise between defense of welfare and a liberal international economic order that had sustained three decades of growth and progress was now seriously destabilized (Cox with Schechter 2002). The golden years were over.

The story of the postwar model of economic development accumulation strategy and its eventual exhaustion starts with the 1970s economic crisis and following recession. However, the post–World War II settlement had shown sign of serious problems as early as the 1960s, with declining profits and the movement of industries to the less developed countries, particularly Asia. The net result was the 1973 recession that shook the capitalist world, leaving it vulnerable to more than two decades of subsequent economic restructuring and social and political readjustment. An epochal change was under way.

However, neither the form nor the content of the restructuring were, or indeed are, taken as given. Instead, these cataclysmic events opened up the terrain to new struggles between social forces—in this case between neoliberals and Keynesians liberals. Even before the crash, "a minority of ultra-liberal economic theologians" (Hobsbawm 1994, 409) had attacked the domination of Keynesian thinking, promoting instead the unrestricted free market as the model of economic development. The attack was also directed at what was regarded as increasingly unruly labor, protected by the entrenched interests of unions.

By 1974, neoliberals were on the offensive, though they did not come to dominate government policy until the 1980s. While seemingly a spontaneous emergence, economists like the Viennese Hayek and Chicago-based Milton Friedman had spent a considerable amount of time since 1947 critiquing welfare-based democracies. As Hobsbawm (1994, 409) observes, "The battle between Keynesians and neoliberals was neither a purely technical confrontation between professional economists, nor a search for ways of dealing with novel and troubling economic problems. . . . It was a war of incompatible ideologies."

Chile was the first testing ground for this new model of economic coordination. Following the ousting of Salvadore Allende's socialist government in a bloody coup in 1973, a pure neoliberal experiment was put into place: privatization of all publicly owned resources (aside from copper), liberalization of finance and openness to Foreign Direct Investment (FDI), freer conditions for trade for firms, and state withdrawal from many social policy programs. This "experiment," however, ended in crisis in 1982, replaced by a more pragmatic implementation of neoliberal theory.

From the 1980s onward all forms of Keynesian policy were purged by international finance organizations, lending agencies, and national governments. Throughout the 1980s neoliberal policies, under the structural adjustment programs of the International Monetary Fund (IMF) and World Bank, were imposed on developing countries in Latin America and sub-Saharan Africa, which were reeling from the worst economic recession since the 1930s. Similarly, in the developed world, neoliberal policies were embraced by political parties of the right and the left, as in New Zealand and Australia, when faced with mounting external debts and rapid inflation following application of Keynesian economic policies.

Vignette 2.1

From Carrie Lips, "The New Trend in Education: For Profit Schools," CATO Institute, November 29, 2000, http://www.cato.org/pub_display .php?pub_id=4437 (accessed August 12, 2007):

> Increasingly, entrepreneurs recognize that the public's dissatisfaction with one-size-fits-all schools is more than just fodder for political debates. It is a tremendous business opportunity.

Restructuring plans imposed by the World Bank on many developing countries and by governments in many developed countries feature key central principles: deregulation, competitiveness, and privatization (Cox 1996, 31). Deregulation refers to the removal of the state from a substantive role in the economy, except as a guarantor of the free movement of capital and profits. Competitiveness is the justification for the dismantling of procedural state bureaucracies and range of welfare provision that were built up in the postwar period. Privatization describes the sale of government businesses, agencies, or services to private owners, where accountability for efficiency is to profit-oriented shareholders. These principles, implemented with the idea of neoliberalism's inevitability, captured in the slogan "There is no alternative," were sold as short-term pain for long-term gain (Kelsey 1993, 10). Labor unions and their activities were strictly curtailed, progressive social policies were restructured to reflect market values, taxation systems were revised to favor the wealthy and ruling classes and business (or in economic terms, capital), many state utilities were privatized, and systems of regulation (particularly around finance) were liberalized (Jessop 2002; Tickell and Peck 2005).

Neoliberalism also legitimated opening up new spaces and means for profit-making, as well as generating the means for the exploitation of labor, for nothing can be allowed to exist outside of the market, including education, knowledge production, and our brains. A profound shift in social and political life emerged, which came to be known as the "Washington Consensus," an idea frequently linked to globalization.

Neoliberals and Their Allies

The early success of neoliberalism can also be attributed to the weakness and confusion of the left due to factors both internal and external. Neoliberalism's ascendancy has mirrored the left's inability to propose a vision of social change, of socialism as a possible alternative to neoliberalism, and of the labor movement as a means to deliver social change to a range of social groups (Waterman 2001). Neoliberalism also found sufficiently broad appeal—particularly in the idea of individual freedom—enabling it to take hold. Neoliberalism was able to resonate with a range of interests, discourses, and agendas within civil society that had been submerged in the postwar class compromise. In the United States, for example, neoliberal ideology spoke to those groups and communities, like the Christian Right, liberal feminists, and black communities, whose identities and projects had been previously denied by the largely white male class project (Apple 2001; 2006). The discourse of "rights" was also mobilized to support opening up previously state-dominated spheres to other actors, offering the very real possibility of setting up new institutional structures, like charter schools or city academies, using a market-based rationale.

Neoliberal policies also resonated among the ruling classes. To be sure, the postwar welfare state in the UK and the United States, designed to redistribute some wealth, impinged on the ruling classes, however growth within the economy in the postwar period, and the specter of socialism as a possible alternative, acted to shore up support among the ruling classes for the postwar settlement. The 1970s crisis of accumulation, however, affected everyone, including the ruling classes. Harvey (2005) argues that when growth collapsed, the upper classes moved decisively to protect their interests, politically and economically. Neoliberalism was the perfect economic engine and political vehicle for this project.

Neoliberalism and Its Contradictions:
From the Washington to Post-Washington Consensus

Neoliberalism, however, had its own internal problems. Its brutal assault on the lives of workers and families—for example, the driving down of wages through vicious suppression of unionization in the developing world (Dale and Robertson 2004)—created problems of considerable social dislocation and social unrest for neoliberal state regimes and associated global agencies and projects such as the World Bank, IMF, and Multilateral Agreement on Investment (MAI). Social movements brought the struggle over the globalization of neoliberalism to the world's attention, as in Seattle.

Toward the end of the 1990s, dogmatic market fundamentalism was undermined in the centers of neoliberal orthodoxy—the UK and the United States. Just as Thatcher and Reagan had played important roles in shaping and transmitting the Washington Consensus, under Clinton, elected in 1992 and Blair in 1997, neoliberalism underwent a transformation. In the World Bank, Chief Economist Joseph Stiglitz in 1997 called this new phase, the "Post-Washington Consensus" (Stiglitz 2002).

However, as Fine (2001) argues, this was not a rejection of the broad trajectory of neoliberal economic policy but rather its deepening and widening. If the 1980s represented a period where the dominant focus was on markets, then the late 1990s can be seen as a return to the social, but always with a focus on the primacy of markets. That is, neoliberalism began to use the language of improving social welfare, of advancing democracy, to explain its economic policies. As one prominent neoliberal economist describes it,

> The Washington consensus advocated use of a small set of [economic] instruments . . . to achieve a relatively narrow goal (economic growth). The post-Washington consensus recognizes both that a broader set of instruments is necessary and that our goals are also much broader. We seek increases in living standards—including improved health and education—not just increases in measured GDP. We seek sustainable development, which includes preserving natural resources and maintaining a healthy environment. We seek equitable development, which ensures that all groups in society, not just those at the top, enjoy the fruits of development. And we seek democratic development, in which citizens participate in a variety of ways in making the decisions that affect their lives. (Stiglitz 1998, 30)

A key policy initiative during the later 1990s centered on the concept of social capital (Fine 2001), an idea central to policies of the "Post Washington Consensus," which is the name given to the form of neoliberalism that we see today. According to the World Bank, "Social capital refers to the internal social and cultural coherence of society, the norms and values that govern interactions among people and the institutions in which they are embedded. Social capital is the glue that holds societies together and without which there can be no economic growth or human well-being. Without social capital, society at large will collapse, and today's world represents some very sad examples of this" (cited in Fine 2001, 158)

Another way to describe the shift in neoliberal policy is that the creation of social cohesion through enhanced social capital was a vital antidote to social instability, seen in increasing social fragmentation, civil conflict, and destabilization produced by migration, immigration, market volatility, and widespread economic and social exclusion.

Vignette 2.2

From the World Bank report, "Global Economic Prospects 2007: Managing the Next Wave of Globalization," December 13, 2006, http://siteresources.worldbank.org/INTGEP2007/Resources/GEP_07_Overview.pdf (accessed August 12, 2007):

> Globalization is likely to bring benefits to many. *By 2030*, 1.2 billion people in developing countries—*15 percent* of the world population—will belong to the "global middle class," up from 400 million today. This group will have a purchasing power of between $4,000 and $17,000 per capita, and will enjoy access to international travel, purchase automobiles and other advanced consumer durables, attain international levels of education, and play a major role in shaping policies and institutions in their own countries and the world economy (emphasis added).

Neoliberalism as a Flawed Model

The work of Karl Polanyi (1944) provides help in understanding why market liberalism, the ideological grandparent of neoliberalism, is so utterly flawed as a way of organizing economies and societies. In his book *The Great Transformation*, Polanyi critiques the work of market liberals like Hayek. In reflecting on why a period of relative stability was followed by fascism in the 20th century, Polanyi argues that the emergence of market liberalism—the idea that markets are self-regulating—emerged as a means of managing the problems of industrialization. This directly led to the Depression, fascism, and all that followed. Market liberalism is based on the view that markets are self-regulating and that markets operate separately from and above or outside society. However, Polanyi argues that markets have always been embedded in social and political life, and the goal of a fully self-regulating market that is disembedded from all of politics and social life is a utopian project. It cannot exist.

In pursuit of this project, however, market liberals must turn human beings and nature into products to be bought and sold—that is, they must commodify them, for instance by patenting medicinal plants that have been used in tribal societies for generations. However, Polanyi argues, market liberalism is based on a lie; firstly, it is wrong or unethical to treat nature and human beings as commodities whose price is determined by the market; secondly, despite the argument that the market is self-regulating, it is evident that the state must play an active role in managing the fictitious commodities of land and labor, as well as important markets.

Polanyi's (1944) extreme skepticism about market liberalism's aims gives rise to his theory that the "free market" requires, despite itself, an alternative movement to stabilize it (the state, and more recently, civil society through concepts like social capital). Bringing the state back in, however, also throws light on the fundamental contradictions of liberalism and its variations, making it an unviable political project in the long term given its propensity to generate social instability, social polarization and social injustice.

Neoliberalism's Transformation of Education

Education in the developed and developing economies has been transformed by neoliberal policies in several important ways. As we will see, however, these policies have been unevenly implemented and experienced, giving rise to important differences across locales, regions, and countries. Changes have also been resisted by workers and their unions—including teacher unions—in some cases more successfully than others. A number of key principles were deployed in the restructuring of education sectors, which changed the mandate (what the education system should do), forms of capacity (the means through which the mandate can be realized, e.g., fiscal and human resources), and mechanisms of governance of the education sector (that is, the means for coordinating the system).

Education systems were mandated to develop efficient, creative, and problem-solving learners and workers for a globally competitive economy, while teachers were to demonstrate through national and global testing systems that they had had taught their young charges. Financing of the public sector, including education, was reduced. Cuts in the education budget were implemented in a variety of ways (Grootaert 1994) and were more likely to affect females than males in that, when family finances are really tight, males are more likely to be provided with the opportunity to learn (United Nations Development Programme 2005, 24–25). It is important to note that cuts in the welfare benefit base, as well as to health and housing, also impact on education so that any assessment of the affect of neoliberal policies on education should look at social welfare policies more generally in order to discern the indirect as well as the direct effects.

Education providers were not only placed under pressure to use funds more efficiently, but they were encouraged to seek additional sources of funding from local and international households (fees for educational services that had previously been free, including full fees from foreign fee-paying students), the business sector (for instance through local donations, direct funding of school infrastructures, public-private partnerships), and from marketing their own services (including curriculum expertise).

The basis for determining a teacher's salary, in many countries part of a system of collective bargaining, was also changed. Teacher unions have been placed under enormous pressure to yield to performance-based or "merit" pay, while governments have used new governance arrangements, such as charter schools (United States), City Technologies (UK), and City Academies (UK) to offer differential wages to teachers. Researchers have noted that in these deregulated U.S. schools teachers were particularly concerned over their right to raise complaints and resolve problems, job security, and levels of pay, and internationally teachers' salaries were seriously affected by austerity measures during the 1980s (ILO 1996). Bonal (2002) shows that fiscal austerity had a negative impact on the quality of teaching and teacher attendance, with educators in Latin American countries having to take up a second job in order to survive.

Finally, new governance systems have been put into place, which reflects a new model for managing the public sector. This involves a number of elements: funding is based on outcomes, some services are decentralized while others are centralized, and departments and institutions have been set into competition with each other through various choice schemes in order to emulate a market (Gewirtz et al. 1995; Ball 2003). Education providers are compared with each other, for instance in the UK through league tables and in the United States with test scores required under No Child Left Behind legislation. A "private" element has been introduced into education. Initially this took the form of privatizing inspection and auditing; outsourcing custodial, catering, testing, counseling and management services; and the establishment of new kinds of education providers—charter schools (Canada and the United States), City Technology Colleges and City Academies (UK), and language schools (Australia). However, more recently, the private, for-profit sector has made even more significant inroads into education. In the United States many charter schools are run by for-profit firms, while services, such as testing and tutoring, make up the growing and highly profitable education industry, complete with rating agencies providing weekly information to investors seeking to manage their "education investment portfolios" (cf. Hentschke 2006). Multinational firms, particularly those involved in the technology industry (for instance Microsoft, Cisco Systems, Sylvan Learning systems) have deepened their involvement in education across the globe, exploring the provision of digitally based (online) education content and provision.

Vignette 2.3

From James Tooley, "Welcome to easyLearn, Class 1," *The Times* (London), April 17, 2006:

> My recent research has shown that between 65 and 75 percent of children in the poorest slums in Africa and India are now in private schools. These schools charge low fees, perhaps a couple of pounds per month. They are run by proprietors who are not heartless businessmen, but who provide free places to orphans and those with widowed mothers. When they tested large random samples of children, my teams found that these schools outperform the government alternative. *And they do it with teachers paid a fraction of the unionised rates.* Unions here would be up in arms about this. Touchingly, the first concern of many delegates at the conferences is that private enterprise would cut teachers' pay or make them work longer hours. But if in the free market, schools can find dedicated champions of children's learning willing to work longer hours, or be flexible on their pay and conditions, then what is wrong with that, if it benefits children?

Vignette 2.4

From a paper by Raymond Majongwe of the Progressive Teachers' Union of Zimbabwe, prepared for the British Zimbabwe Society Open Forum in 2006:

> Conditions of work in the education sector are deplorable. Teachers, School Heads and Education Officers all earn salaries that are below the PDL [Poverty Data Line]. On average a teacher takes home a monthly salary of $25,000 against a PDL pegged at $96,000 per family of six. Teachers are no longer creditworthy and cannot purchase, even through installment schemes, items like TV sets and refrigerators.

Who Wins and Who Loses from Neoliberal Policies and Programs?

Has neoliberalism managed as a development model, as its advocates contend, to deliver long-term gain despite short-term pain? And if so, is gain widely distributed—based on the liberal market thesis that the gains trickle down? Or, have the spoils been largely grabbed by a small class fraction? In this section I will begin by reviewing the bigger evidence on this question, before focusing specifically on the education sector.

Solid evidence is available that neoliberalism's globalization has not delivered greater equality and reduced poverty across the globe, despite earlier claims by the World Bank and neoliberal economists. A series of reports (ILO 2004; UN 2005; UNDP 2005) reviewing the affects of economic globalization point to a slowdown in economic growth, increased levels of inequality and poverty, and the marginalization of the "fourth world"—Sub-Saharan Africa. The countries that display the greatest levels of growth and decreased levels of poverty are China and India, both of which did not follow neoliberal economic policies. Indeed, when China and India are removed from analyses, it is evident that there is an overall *decline in the developed economies, at the same time featuring an increase in levels of inequality within a significant number of developed countries* (Robertson et al., 2006). Declining income has had a major affect on levels of participation in education—particularly in Sub-Saharan Africa.

What have been the effects on education specifically? Overall, levels of expenditure on education decreased rather than increased over the 1980s and 1990s, with the exception of Latin America. However the remaining question is whether this decline was felt more generally, or whether some classes or class fractions were able to either protect themselves from, or benefit from, the state's neoliberal policies and programs. Brint (2006, 185) provides good evidence that in countries such as the United States, the age of increasing educational opportunity ended around the time of the implementation of neoliberalism; that is, the early 1980s. Since the 1980s, the United States has made fewer efforts to foster the socioeconomic mobility of children from lower classes and minority groups through education. Most notably, the public resources to support equality of opportunity at the college level have declined: "At the leading private universities, tuitions are more than half the average family's total yearly income. Without scholarship aid, children whose families are not in the top few percent of income cannot afford to attend these colleges, no matter how well qualified they may be. Even at public universities, fees have risen at nearly three times the rate of inflation since the 1980s, much faster than the incomes of all but the very rich" (Brint 2006, 185).

Similarly, in the UK, choice policies have tended to favor the middle and ruling classes, who are able to use their social, economic, and cultural capital in order to secure an education at a private school or at one of the high-status publicly funded comprehensive schools, while conversely, these schools can "cream" the "best" (largely middle- and upper class) students, thus reinforcing

> **Vignette 2.5**
>
> Extract from a UK business, School Select, offering to help those parents who can afford a large sum of money to bypass schools' admissions regulations and get their child into the school of their choice. "School Select—Professional Fees," http://www.schoolselect.co.uk/fees.html (accessed August 12, 2007):
>
> As with most things in life, there is a general price–quality equation; other web providers offer admission appeals support from as little as £5. Please be assured our service is ultra professional and your assignment will be handled from start to finish by a former UK Head Teacher. . . . Typically a School Search report will cost in the region of £595 with accompanied visits being charged at approximately £645 per day depending on the distance traveled. . . . Professional fees for Admission Appeals work are as follows—the preparation of your case costs £745; representation at the Hearing costs £895 plus reasonable travelling expenses. . . . Other assignments are charged for at our standard hourly rate, currently £95.

class divisions and relations (Robertson and Lauder 2001). Poorer families as a result must accept neighborhood schools or under-chosen schools while being blamed for their poor performance (Whitty et al., 1998). While these schools are not necessarily "bad," they have low status, which closes off access to further privileges and both mirrors and produces the system of social stratification. Where the state has stepped in to improve schooling for poor families, research has found that these schemes have been dominated by the middle class (Edwards et al., 1989). The picture is much the same in countries like New Zealand, where radical neoliberal policies were implemented in the education sector throughout the 1990s. Lauder and Hughes (1999) report a spiral of decline in working class schools, leaving teachers and students at risk of low morale, low funding, and poor performance.

Transformations of cities have also had an impact on the lower classes, who in many cases are being driven out of the big European cities, such as Paris, because of the rise in housing prices and the development of business districts (Van Zanten 2005, 158). This results in concentrations "of lower class and immigrant children in neighborhoods and schools of the urban periphery, which in turn leads to the kinds of socialization inside and outside of schools that can be characterized as 'peripheral' in more than a geographic sense" (ibid, 159).

In looking further afield to countries that have not implemented neoliberal policies in quite the extreme way that we see in the UK, the United States, and New Zealand, it is possible to see that class inequalities are diminished through schooling in Sweden and the Netherlands. Although redistribution policies are under attack in these countries, values of social equality for the community rather than equal opportunities for individuals still dominate (Brint 2006, 182–83).

A recent study on markets and governance across five European countries (Belgium, France, Portugal, England, and Hungary) by Ball and his colleagues (cf. Maroy, forthcoming) discerns quite different engagement with, and resistance to, neoliberal education policies, thus producing a different mode or style of neoliberalism. In the case of France, for instance, neoliberal ideas did not possess strong organizational bases (in contrast with Britain and the United States)—there was no equivalent of the think tanks, press, and financial sector. This resulted in France engaging with neoliberal policies in a more pragmatic way than they were engaged in Britain and Chile, where it was a highly political process. Nevertheless, it is still possible to see the broad contours of a neoliberal settlement in France—the role of markets in economic regulation, the promotion of free trade of goods and capital, and the prioritization of the fight against inflation.

Larry Kuehn explains in his essay how teacher unions have mediated neoliberalism—a conclusion that I also drew in my own analysis of the history of teachers' work in the United States compared with the UK (Robertson 2000). Indeed, unions work as a form of social capital, providing a network of resources that enable teachers to protect themselves from overt forms of exploitation. It is no accident that in order to discipline and exploit teachers as workers, neoliberal policies attack teacher unions (Robertson 2003). This of course has not been a straightforward process, though as Waterman (2001) points out, many labor unions have been spectacularly poor at thinking through what it means to be a union in the context of globalization.

Despite their limitations, labor unions have been successful in stalling the onward march of neoliberalism in some locations. Jones (2005) shows that the political traditions associated with the postwar reforms in Italy, France, and England meant that it was possible to mobilize power and mount a concerted campaign of action opposing the neoliberal shaping of education. Scholarship on Latin America illuminates how implementation of neoliberal reforms partly reflected differences in the intensity of labor's opposition to the reforms, the internal dynamics of teacher unions, and the contradiction between the rights of teachers as workers and their view of themselves as professionals, as well as an ambivalence over processes of centralization versus decentralization (Madrid 2003).

Where To from Here

It is tempting to think that neoliberalism as a political and class project might implode because of its own internal contradictions of offering freedom but instead tightening the shackles of control and of commodifying all in its wake and yet needing to legitimate itself as an ideology that has something for everyone. The evidence of these contradictions is clearer and clearer. Neoliberalism might be seen as having caused teachers to rethink what we do and how, by challenging fundamental assumptions about the purposes of education, it might thus be seen as having value as a social and individual good. But it is clear that in sum, the working and middle classes have paid a huge price for reforms in schooling carried out in accordance with neoliberal ideology. And while, for the moment, we can point to big and small struggles and victories, my informed hunch as a researcher is that the battle has barely begun.

The contributions to this volume provide compelling evidence that education has all too silently been rapidly commercialized and is becoming big, big business, protected by global regulations. Pressure mounts on willing national governments (for example from the OECD and World Bank) to cut their losses and stop attempting to transform archaic, bureaucratic, and difficult education systems and their teachers, and instead go with the latest technological solution. Personalized learning is the new buzzword, a learning experience ordered over the Internet and packaged up "just for me" (Organisation for Economic Co-Operation and Development 2006). International agencies promote these as solutions to a myriad of problems facing education policy makers, from providing universal access to children in Africa to promoting nations to become winners in the global knowledge economy. These developments are being protected by the legally institutionalized rights of capital—the new investors in education—in agreements being negotiated as part of the WTO's General Agreement on Trade in Services (see Robertson et al., 2002).

Yet, as contributions to this book show, alternatives are emerging that give us a way to think differently about educational improvement, as well as paths to get there. This means, however, confronting face on what has happened and why. It means daring to talk about the worsening conditions of teachers' work and identifying the changes as increased exploitation and not improved performance. It also means talking in schools and unions about the transformation of teachers' work and workplaces as the outcome of a class project.

References

Apple, M. 2001. Comparing neoliberal projects and inequality in education, *Comparative Education* 37 (4): 409–23.

———. 2006. Understanding and interrupting neoliberalism and neoconservativism, *Pedagogies: An International Journal* 1 (12): 21–26.

Ball, S. 2003. *Class strategies and the education market: The middle classes and social advantage*, London and New York: Routledge.

Bonal, X. 2002. Plus ça change . . . The World Bank global education policy and the Post-Washington Consensus, *International Studies in Sociology of Education* 12 (1): 3–22.

Bourdieu, P. 1998. Utopia of endless exploitation: The essence of neoliberalism. *Le Monde Diplomatique* (December). http://mondediplo.com/1998/12/08bourdieu. Accessed November 29, 2007.

Brenner, N. 2004. *New state spaces: Urban governance and the rescaling of statehood*. Oxford: Oxford University Press.

Brint, S. 2006. *Schools and societies* (2nd ed.). Stanford, CA: Stanford University Press.

Cox, R. 1996. *Approaches to world order*, Cambridge: Cambridge University Press.

Cox, R., with M. Schechter. 2002. *The political economy of a plural world: Critical reflections on power, morals and civilization*. London: Routledge.

Dale, R., and S. Robertson. 2004. Interview with Robert Cox. *Globalization, Societies and Education* 2 (2): 147–60.

Edwards, T., J. Fitz, and G. Whitty. 1989. *The state and private education: An evaluation of the assisted places scheme*, Lewes: Falmer.

Fine, B. 2001. *Social capital versus social theory: Political economy and social science at the turn of the millennium*. London: Routledge.

Fourcade-Gourinchas, M. and S. Babb. 2002. The rebirth of the liberal creed: paths to neoliberalism in four countries, *American Journal of Sociology*, 108 (3): 533–79.

Gewirtz, S., S. Ball, and R. Bowe. 1995. *Markets, choice and equity*, Buckingham: Open University Press.

Grootaert, C. 1994. Education, poverty, and structural change in Africa—Lessons from Cote-d'Ivoire, *International Journal of Educational Development* 14(2): 131–42.

Harvey, D. 1989. *The condition of postmodernity*, Oxford: Blackwells.

———. 2005. *A brief history of neoliberalism*, Oxford: Oxford University Press.

———. 2006. *Spaces of global capitalism: Toward a theory of uneven geographical development*, New York: Verso.

Hentschke, G. 2005. *Characteristics of growth in the education industry: Illustrations from US Education Businesses*. Paper presented to the symposium on New Arenas of Educational Governance: The Impact of International Organizations and Markets on Educational Policymaking, September 23–24, at the University of Bremen, Germany.

Hobsbawm, E. 1994. *Age of extremes: The short twentieth century 1914–91*, London: Abacus.

Hood, C. 1995. The "New Public Management" in the 1980s: Variations on a theme. *Accounting, Organizations and Society*, 20 (2/3): 93–109.

International Labor Organization. 1996. *Impact of structural adjustment on the employment and training of teachers*. Report for discussion at the Joint Meeting on the Impact of Structural Adjustment on Educational Personnel, in Geneva, Switzerland.

———. 2004. *A fair globalization: Creating opportunities for all*. Geneva: ILO World Commission on the Social Dimensions of Globalization.

Jessop, B. 2002. Globalization and the national state. In *Paradigm lost: State theory reconsidered*, ed. S. Aronowitz and P. Bratsis, 185–220. Minneapolis: University of Minnesota Press.

Jones, K. 2005. Remaking education in Western Europe. *European Educational Research Journal* 4 (3): 228–42.

Kelsey, J. 1993. *Rolling back the state*, Wellington: Brigit Williams Books.

Lauder, H., and D. Hughes. 1999. *Trading in futures: Why markets in education don't work*. Basingstoke: Open University Press.

Macpherson, C. B. 1962. *The political theory of possessive individualism: Hobbes to Locke*. Oxford: Clarendon.

Madrid, R. 2003. Laboring against neoliberalism: Unions and patterns of reform in Latin America. *Journal of Latin American Studies* 35:53–88.

Maroy, C. (forthcoming). Review of "Convergences and hybridization of educational policies around post-bureaucratic models of regulation." *Compare*.

Molnar, A. 2006. The commercial transformation of public education. *Journal of Education Policy* 21 (5): 621–40.

Ong, A. 2006. *Neoliberalism as exception: Mutations in citizenship and sovereignty*. Durham, NC: Duke University Press.

Organisation for Economic Co-Operation and Development. 2006. *Personalising education*. Paris: Organisation for Economic Co-Operation and Development.

Polanyi, K. 1944. *The great transformation: The political and economic origins of our time*. Boston: Beacon Press.

Robertson, S. 2000. *A class act: Changing teachers' work, the state and globalisation*, London: Falmer.

Robertson, S. 2003. The politics of re-territorialization: Space, scale and teachers as a profes-sional class. In *Teachers and European integration*, ed. H. Athanasiades and A. Patramanis, 41–69. Athens, Greece: Educational Institute.

Robertson, S. and H. Lauder. 2001. Restructuring the education/social class relation: A class choice. In *Education reform and the state: Twenty-five years of politics, policy and practice*, ed. R. Phillips and J. Furlong, 222–36. London: Routledge/Falmer.

Robertson, S., X. Bonal, and R. Dale. 2002. GATS and the education services industry. *Comparative Education Review* 46 (4): 472–96.

Robertson, S., M. Novelli, R. Dale, L. Tikly, H. Dachi, and A. Ndebela. 2006. *Education and development in a global era: Ideas, actors and dynamics in the global governance of education*, Report to the DfID, London.

Stiglitz, J. E. 1998. Towards a new paradigm for development: Strategies, policies and processes. Prebisch Lecture for UNCTAD, October 19, in Geneva, Switzerland.

———. 2002. *Globalization and its discontents*. New York: Penguin.

Tickell, A., and J. Peck. 2005. Making global rules; Globalization or Neoliberalism? In *Remaking the global economy: Economic-geographical perspectives*, ed. J. Peck and H. Yeung. London: Sage, 163–81.

United Nations Department of Economic and Social Affairs. 2005. *The inequality predicament: Report on the world social situation 2005*. http://www.un.org/esa/socdev/rwss/rwss.htm (accessed January 7, 2006).

United Nations Development Programme. 2005. *Human Development Report 2005: International Cooperation at a Crossroads: Aid Trade and Security in an Unequal World*. http://hdr.undp.org/reports/global/2005 (accessed January 7, 2006).

Van Zanten, A. 2005. New modes of reproducing social inequality in education: The changing role of parents, teachers, schools and educational policies. *European Educational Research Journal* 4 (3): 155–69.

Waterman, P. 2001. Trade union internationalism in the age of Seattle *Antipode* 33 (3): 312–26.

Whitty, G., S. Power, and Halpin, D. 1998. *Devolution and choice in education*. Basingstoke: Open University Press.

PART II

Neoliberalism's Global Footprint

CHAPTER 3

Education Reform under Strangulation

John Nyambe

Education reform has been placed high on the agenda ever since Namibia's independence from apartheid South Africa in 1990. But in recent years I have begun to ask, to what extent does Namibia shape its own educational destiny via its own national bodies? Does the fate of an education system, such as that of Namibia, lie in the hands of the transnational policy formulating agencies such as the World Bank?

Over the past fifteen years of Namibia's independence, I have actively been participating in the reform of Namibian education. Immediately after independence, I was appointed as a teacher educator (lecturer) at one of Namibia's four colleges of education, Caprivi College of Education in the far northeast of Namibia. Five years later, I was appointed as an educational leader (vice-rector) at another college of education (Ongwediva College of Education) in the far northern part of Namibia. In 2001 I returned to Caprivi College of Education in yet another leadership position as the Rector of the College, the position I held at the time of writing this narrative.

Through my active involvement in facilitating classes; providing leadership; participating in curriculum meetings, workshops, and conferences; as well as occasionally writing in the *Reform Forum* (a journal for educational reform in Namibia), I have not only come to experience the confusion, tensions, and frustrations associated with the current transformation in Namibian education, and teacher education in particular, but I have also come to grapple with some unresolved questions related to the whole reform process, especially of late.

One such persistent question has been the question of whether national bodies such as the National Institute for Educational Development (NIED), in the Namibian case, or even the Ministry of Education, for that matter, have much relevance in the face of globalization and the transnational policy-formulating agencies such as the World Bank or USAID (United States Agency for International Development). How much of the educational reform that we have witnessed and continue to witness have come about as part and parcel of the global dictates of the agencies of policy formulation such as the World Bank and regional groups like it?

In particular, I refer to the most recent but very extensive educational reform program in Namibia, encapsulated as the *Education and Training Sector Improvement Programme* (ETSIP), a sector-wide reform program crafted following a study in 2003 by the World Bank. The World Bank study, *Human Capital Development and Knowledge Management for Economic Growth with Equity*, assessed the adequacy or readiness of Namibia's education and training system to effectively support the attainment of national development goals, especially the aspired-for transition to a knowledge economy (World Bank 2003). Subsequently, eight sub-studies, coordinated by the World Bank and augmented with its technical advisors, have been produced on the same subject.

Based on the World Bank's reports and technical advice, ETSIP, a ten-year strategic plan for the Ministry of Education in Namibia, was produced. Hence, at the forefront of the ETSIP document the Ministry of Education acknowledges with "gratitude the substantive technical support and guidance provided by the World Bank" (Government of the Republic of Namibia 2005, ii)

Namibia is embarking upon the biggest reform ever to cover the entire education sector, with the priorities closely reflecting the recommendations of the World Bank (2003) report. Central to the recommendations of the World Bank is the persistent advice to cut public expenditure on education and to increase "private sector provision of education and training," which the Bank sees as "critical for future expansion of access, especially at the secondary school level" (World Bank 2003, 106). This recommendation radically contradicts the Namibian policy of *Towards Education for All* (Ministry of Education 1993). Instead of facilitating the implementation of the Education for All policy, the recommendation seems to cause Namibia to revert to elite education. This is so because in the setting of a privately owned

Vignette 3.1

Burkina Faso, in West Africa, is one of the poorest countries in the world. Primary school enrollment was at 36% in 2004. Parents have to pay school fees. It is at present receiving the attentions of an army of consultants as part of the Structural Adjustment Plan from the IMF. Excerpt from a study issued by Education International, "Effects of Structural Adjustment Policies in Burkina Faso."

In its reports, the SNESS underlines that the rule which limits class size to 70 students per class in the first 'cycle' and 60 students in the second 'cycle' is generally disregarded, especially in the two largest cities, Ouagadougou and Bobo-Dioulasso. According to the SYNTER, class sizes are continually on the rise: over 130 pupils in primary education and 120 at secondary level are not uncommon. At the university, up to 1,000 students crowd into lecture halls with a seating capacity of 600.

school, the ultimate responsibility is not to children, students, or the community but to the owners and shareholders of the enterprise.

In the quest to maximize shareholders' capital, school fees are increased but become unaffordable to the ordinary citizen. The ultimate responsibility is to maximize shareholders' returns and not necessarily to advance the public good. Only the wealthier families will afford to send their children to such private schools, a situation that leads to social stratification and elite formation. This situation militates against the "access" (MEC 1993, 32) that is one of the major goals of the Namibian education system.

In addition to privatization, the Bank further recommends that apart from governmental public spending on education, "loans and grants from donors and international organizations" be another method to raise the required capital investment in education (World Bank 2003, 106). Yet, the literature, as well as experiences elsewhere on the African continent, abounds with evidence of the brutal consequences, mainly to the marginalized poor, of loan or aid money. The structural adjustment programs that come along with the repayment of the loans have severe negative consequences for the marginalized majorities.

In addition to this, the Bank draws on its analyses of the Southern Africa Development Community (SADC), middle-income countries that seem to record higher education effectiveness for much lower spending on education. The World Bank advises Namibia that "the physical capacity of existing classrooms and other teaching facilities can be substantially and immediately expanded by increasing the number of schools on double shift and by increasing

> **Vignette 3.2**
>
> From Sudha Ramachandran, "Literacy Beats Out Education in India," *Asia Times,* April 24 2004:
>
> Under World Bank pressure, the government is moving away from earlier commitments that it had made on education. In the National Policy on Education 1992, the government committed itself to providing three teachers per primary school. Under the World Bank–sponsored DPEP, an "innovative scheme" of multi-grade teaching was introduced. This allows a single teacher to handle five classes simultaneously.

class size from the current average of 33.5 to 40 learners" (World Bank 2003, 105).

The World Bank's proposal to cut public expenditure on education and increase class sizes occurs as the Namibian Government has been battling with the large class size in the northern parts of the country, where class sizes ranging from forty to fifty or even more have posed a serious problem to the implementation of learner-centered pedagogy. Bigger class sizes have proved barriers to effectively implement learner-centered pedagogy, the approach that is widely advocated by the Namibian education system.

It is worth observing that not only is this recommendation contradictory to the Namibian learner-centered education, widely acclaimed as the philosophical underpinning of the Namibian education system, but it also militates against effective learning. The motive underpinning the World Bank's recommendations seems to be above all to cut public spending regardless of whether or not this undermines the quality of education. The Bank's recommendations support Hill's (2003) argument that national and global capitalism seek to cut public expenditure because public services are expensive, a tax on capital. Cuts in public expenditure serve to reduce taxes on profits, which in turn increases profits from capital accumulation.

Further recommendations by the Bank have included, among others, the formulation of "National Professional Standards for Teachers," the introduction of a five-year Teacher Licensing System as well as the contemplated Performance Contract System for school managers. These measures are proposed by the World Bank with a rationale to strengthen the quality, effectiveness, and efficiency of the general education and training system. However, it is evident that in Namibia, as elsewhere in the world, measures undertaken in the name of making teachers more accountable are an attempt to heighten control, predictability, and deprofessionalization. No longer will teachers exercise their own professional judgment, as their professional behaviors will be dictated by the professional standards to which they have to adhere if they are to secure their teaching licenses and chances of being hired as teachers.

It is, therefore, evident from the foregoing discussion that not only do these global and globalizing forces maintain their stranglehold on the education systems of countries such as Namibia but they also stifle local expertise and oversight of their own education system's destiny.

The global interest of corporations and transnational agencies in education reform is prompted, among other factors, by the fact that the education sector has a special role in shaping, developing, and molding the future labor power of the capitalist system. The education sector is responsible for the social production of labor power; equipping students with skills, competencies, abilities, knowledge and attitudes, and personal qualities needed by the capitalist system.

The education sector, and the teacher within it, therefore takes a guardian role over the quality and attitudes of labor power, hence the Bank urges measures such as the development of "professional standards," "competencies," and new licensing methods to ensure that proper control of teachers is maintained. Mechanisms for teacher regulation of their own pedagogies are put in place to ensure that labor with appropriate attitudes is produced in schools.

Another aspect of curriculum control by the transnational policy formulating agencies is ensuring that modes of pedagogy that are antithetical to labor power production do not find their way into schools. Transnational agencies seek to stifle any forms of pedagogy that attempt to educate students and raise their consciousness regarding their real position as future labor power and underpin such critical consciousness with critical insight that seeks to undermine the social production of labor.

I would like to conclude by stating that over the fifteen years that I have been actively involved in the reform of the Namibian education, I have come to realize that local government agencies such as national curriculum bodies have very little relevance, as much of curriculum reform increasingly the mandate of transnational organizations. However, these transnational policy formulating agencies are not driven in their activities to maintain the public good. Instead, they are driven and propelled by the insatiable demand for profit. In policy formulation, profit, instead of public welfare, occupies the center stage.

References

Alexander, T., E. Norrby, and U. Teichler. 2004. *Building the foundations for a knowledge-based economy and society: A review of tertiary education in Namibia.* Washington, D.C.: World Bank.

Government of the Republic of Namibia. 2005. Education and training sector improvement programme (ETSIP). Windhoek: Ministry of Education.

Hill, D. 2003. Global neoliberalism: The deformation of education and resistance. http://www.jceps.com.index.php?pageID=article&articleID=7 (accessed November 17, 2003).

Ministry of Education and Culture. 1993. Towards education for all: A development brief for education, culture and training. Windhoek: Gamsberg, Macmillan.

Ministry of Education 2006. Draft national standards for teachers. Windhoek: Ministry of Education.

Marope, M. 2003. Namibia human capital and knowledge development for economic growth with equity. Country Department: AFC01, Africa Region. Washington, D.C.: World Bank.

CHAPTER 4

Teaching for the Factory
Neoliberalism in Mexican Education

Rodolfo Rincones,
Elaine Hampton, and Cesar Silva

Motivated by the neoliberal impetus of the 1980s, the Mexican government initiated several reforms in its economic and social sectors, which in 1992 reached the education sector. Mexican education had become a bastion of pride and nationalism because of phenomenal reform and improvements during the mid-twentieth century.

With Mexico's 1994 inclusion in the Organization for Economic Cooperation and Development (OECD) and the signing of the North America Free Trade Agreement (NAFTA), the Mexican government purposefully privileged the alignment of public education funds with local or regional corporate needs, especially in technical education, providing justification for the growth of two-year technical colleges. Technical education in Mexico is mostly concentrated at the high-school level and above. Several types of schools, depending on geographical location, offer both preliminary courses of instruction for higher education and technical skills and knowledge for those students who do not want to continue their education. Our aim here is to describe what we have learned in our study of changes in curriculum in technical education.

Much of our study focuses on Ciudad Juárez, Chihuahua, México, the sister city of El Paso, Texas. The city hosts about 278 foreign manufacturing plants, mostly serving U.S. corporate needs in areas such as automobile parts and electronic equipment. These factories are called *maquiladoras*. The city is host to the largest population of *maquiladora* employees in México, the vast majority being factory line workers, earning about six dollars a day. Technicians account for about 15 to 20 percent of the employees.

The Conalep Schools

Mexican students can attend several types of high schools, depending on their interest, location, test scores, availability, and cost. One of these is Conalep (*Colegio Nacional de Educación Profesional Técnica*), which was designed in 1979 to prepare citizens to be technical professionals. The course of study is three years in six semesters designed specifically to fit the student to enter business management. Students who wish to attend the university must take a complementary course of study that includes one additional course each semester.

Students in Conalep choose a career focus. The curriculum for the industrial productivity career path in the Conalep we visited in Ciudad Juárez includes the study of production schedules, materials management, inventory control, production and control systems to maximize productivity, and safety in the work environment. The workshop for this industrial productivity program was a mock *maquiladora* assembly line, complete with the yellow line drawn on the floor to direct traffic flow. The table and stools for the assembly line were castoffs from a *maquiladora*. Students practiced assembly line procedures by making wooden boxes from pieces of scrap wood stored on the shelves.

The school is in a community where several *maquiladoras* are located. After graduation, many of these students cross the street and become employees in the factory, working on the assembly line for approximately six dollars per day.

ICATEP in Puebla

The Mexican state of Puebla provides an example of another school-industry collaboration in technical education, secondary training schools called *Instituto de Capacitación para el Trabajo del Estado de Puebla (ICATEP)*. The schools were designed to prepare people to work in industries in the local community, one goal being to provide students in rural areas and indigenous communities with jobs that will keep them in the local area. Students select from a curriculum of courses in carpentry, secretarial skills, sewing, automotive skills, and computer training. The federal government provides funding for the teachers' salaries, the facilities, and some of the basic tools such as the

Vignette 4.1

The new British Prime Minister, Gordon Brown announced to business leaders the formation of the Council for Educational Excellence. Almost all the appointees are business leaders;, there are no teacher union appointees, and Damon Buffini is a Private Equity boss.From "Speech by the Chancellor of the Exchequer, the Rt Hon Gordon Brown MP, to Mansion House," HM Treasury, http:// www.hmtreasury.gov.uk/newsroom–and –speeches/press/2007/press_68_07.cfm (accessed July 7, 2007).

It is good for our country that we have businesses involved in some schools, and I can congratulate companies who are. In future every single secondary school and primary school should have a business partner, and I invite you all to participate. Every secondary school should have a university or college partner, every school should work directly with the arts and cultural and sporting communities in their area, every school should work with other local schools to raise standards for all.

I am pleased that Sir Terry Leahy, Sir John Rose, Richard Lambert, Bob Wigley and Damon Buffini have agreed to join the Council.

The Council will be advised by Sir Michael Barber. Julia Cleverdon, Head of Business in the Community, has agreed to report on how more businesses, small, medium, and large, can play a bigger part in support of our schools.

twenty computer stations and auto mechanics tools, along with a small budget for maintaining the machines. The rest of their budget comes from support from the community, and there is a requirement that ICATEP schools link with the local community for funding and support.

The school is near a small city surrounded by many small rural communities and a large U.S.-owned clothing factory. Industry-school collaboration is most evident in the curriculum for sewing. The clothing factory provided thirty sewing machines on loan and pays tuition for the students. In return, the students work for two hours per day at the factory while they are learning at the school. At the end of the one-semester sewing class, the students had completed one garment. The local police provided fabric, and the women in the sewing class made new police uniform shirts as their learning projects. The owner of the clothing factory said that this program is very beneficial to his business. He can select the best students to fill his vacancies, and the time to train new employees is reduced significantly from seven weeks down to about three weeks.

The ICATEP school provides an opportunity for self-improvement for the women who enter its program—in a geographical region were opportunities are few. During our visit we were caught up in the enthusiasm about this short-term benefit. The school director ensured that the school honor cultural arts and traditions. However, when reflecting on the sewing curriculum, we ache over the women's missing education about democratic thinking and leadership and how to implement them. A democratic education for these women was a far cry from sewing the same seam on the same cloth over and over again for one of the thousands of athletic wear stores in one of the thousands of malls in *los Estados Unidos*.

CBTIS in Tijuana

A large Chinese aviation factory assisted in the creation of a school in Tijuana designed specifically to meet the needs of this one major industry. With the support of a powerful Mexican senator, government and private funds from Mexico and Hong Kong were used to build a technical high school, *Centro de Bachillerato Técnico Industrial y de Servicios*, with a curriculum designed specifically for the aviation industry's needs. During its first few years, most graduates immediately stepped into technical positions at the *maquiladora*. Then, with a downturn in the economy in the 1990s, the *maquiladora* scaled back production, and many employees lost their jobs, leaving new graduates with no options: their training had been specific to this factory's needs. The city was suddenly forced into large expenses to add other fields of study to the school to provide a larger range of employment options.

Universidad Tecnológica

The technical universities, *Universidades Tecnológicas*, were created in 1991 as a strategy to provide technicians whose education would prepare them for the needs of local or regional industries, to increase productivity.

Universidades Tecnológicas are in twenty-eight out of the thirty-two states. Their number increased from forty-eight in 2001 to sixty-one in 2006, a 5 percent growth. In the same period of time, the enrollment in these institutions increased 57 percent—from 42,609 in 2001 to 66,660 in 2006.

Prospective students and parents consider this type of technical higher education attractive because it takes less time to complete a degree, it is less expensive, facilities have state-of the-art equipment, and students who finish the program have better employment opportunities. For instance, in Ciudad Juárez the program of study was certified via international standards of training for its alignment with the *maquiladora* industry, feeding the perception of guaranteed employment of the graduates. This certification and alignment, however, has not favored such employment. Of the 222 graduates at the Cuidad Juárez university, only 101, or 45 percent, are employed. At the national level, official statistics indicate that rates of employment have decreased. In 2001, 69 percent of the students had employment. This percentage decreased to 63 percent in 2005.

Once they are employed, conditions for the majority of these graduates are not any better than those for workers who do not have this university degree. A recent evaluation of the Mexican *Universidades Tecnológicas* indicated that earning this degree does not guarantee employment and that this degree is not recognized by the industry: there is no salary level specific to this type of university graduate. Graduates observe that their salaries are low and similar to salaries of technicians without this degree. Evaluators warn that these facts jeopardize the evolution and existence of these universities.

In addition, employment conditions of faculty at these technical universities can be problematic. Faculty perceive that their jobs are less prestigious that those at traditional four-year universities. As of 2006, the *Universidades Tecnológicas* had nationwide 2,343 full-time faculty members. Of these, 55 percent had bachelor's degrees only, 44 percent master's degrees or specialization, and only 1 percent had doctorates. The employment conditions of these faculty members is also uncertain in that they have to renew contracts every four months.

Conclusion

Technical education based on competencies that are appropriate to the export-oriented private manufacturing sector is a strong priority in Mexico's education system. Multinational corporations have migrated to Mexico and established *maquiladoras* because of the existence of large pools of inexpensive labor and favorable economic policies through tax breaks. Educational curriculum and public educational funds have been diverted to fill the needs of these foreign economic entities.

As teacher educators and researchers, we have serious concerns about the ethics of the curriculum reforms we have documented and changes in curriculum driven by the power of external economic needs. We urge a close examination of the extent of the role of public education in meeting the needs of private industry. The need for scrutiny is exacerbated when the private industry providing the education is foreign to the nation. We encourage technical education, but we would argue against and oppose policies that promote the expenditure of public funds to subsidize the production of these foreign industries.

CHAPTER 5

Neoliberal Education in Denmark

Jette Steensen

Seen from a distance Finland, Sweden, Denmark and Norway may appear more or less similar. Beginning at the end of World War II, an alliance around the goals of the welfare state were established in all Scandinavian countries, and the rules for cooperation between the state, the unions, and employers were integrated with state law. Social Democratic parties governed alone or participated in coalition governments for many years and have contributed in establishing strong welfare states which are difficult to dismantle overnight, especially when accompanied by a general high standard of living. Thus Scandinavia might still appear as the antithesis of neoliberalism. Yet there are differences among the various states that make educational restructuring occur in different shapes and at a different pace.

During the 1990s the Social Democrats slowly introduced neoliberal restructuring in order to prepare for the new international competition created by globalization. When the Social Democrats lost government power—in Denmark in 2001 and in Sweden in 2007—the neoliberal trend accelerated, cleverly disguised in the rhetoric of quality reform of the welfare state. The rhetoric, however, is accompanied in education by a lack of funding, often covered up by the shifting of responsibilities between the centralized and decentralized levels of the public sector, and reforms at all levels of the educational sector from university to kindergarten. To paraphrase Ingrid Carlgren (1994), the process of destruction of the Scandinavian welfare state is like removing the tent pegs without telling anybody the intention is to tear down the tent.

Changes have accelerated since 2001, when (neo)liberal prime minister Anders Fogh Rasmussen formed a coalition government with the Conservative

Party and the extreme right populist and very nationalist Danish People's Party. More clear-cut neoliberal principles have combined with strong state interference in matters previously left to local educational authorities and along somewhat authoritarian lines.

Denmark was always the most liberal Scandinavian state, long known for its permissiveness in all areas of life. It was never heavily industrialized, and it has a strong tradition of independent petit-bourgeoisie and a wealthy independent agricultural sector. These factors have made the society very receptive to individualistic and free-enterprise ideologies as well as populist movements. Secularization in Denmark has not meant de-Christianization of mentalities: cultural radicals who are progressive are minorities. Danish neo-conservatives have forged new ties between the state and established religion, utilizing the populist move against immigrant minorities, whom they consider the cause of all evils, as demonstrated in the so-called Mohammed crisis. The government's defence of a newspaper's publication of cartoons defaming Islam was cunningly marketed for domestic policy purposes as a matter of freedom of speech. In reality the incident continued a pattern in which right wing officials publicly took steps to denigrate members of minority populations, who are used wherever and whenever scapegoats for globalization's adverse effects are needed.

Basically Denmark's "quality reform" in education means reduced public spending, additional evaluation and management, and public bashing of professionals. Key elements include the following:

- Introduction of a national curriculum
- Introduction of tests in grade three, six, and nine
- Linking of school managers' salaries to the performance of the school
- Prioritization of the teaching of mathematics, natural science, and Danish in schools
- Introduction of individualized student plans
- Removal of social sciences from the teacher education curriculum
- Increased ratio of students to teachers
- Closing of a number of small schools

As public sector professionals are officially blamed for the state of the public sector, it has become increasingly difficult to attract qualified teachers, doctors, nurses, social workers, etc. This again leads to an increase in private alternatives (e.g., health insurance that ensures access to special treatment at private hospitals for the middle class). The attacks on schools and teachers have also resulted in a steep decline in the pool of applicants for teacher education. The main solution so far has been the introduction of shorter alternative certification programs. In this way the tradition of four-year teacher education programs is being gradually eroded by the market. Another result has been steadily increasing attendance at private, "free" (parochial) schools: currently around 14 percent of Danish children attend these schools, a relatively high percentage for a country with a strong belief in public schooling. One explanation might be that due to its history of liberalism, Danish free schools have always been considered a very positive supplement to public schools, and they are subsidized 75 percent by the state. Parochial schools are not a hugely expensive alternative for parents and provide a culturally acceptable excuse for fleeing the public school system, a predicament that is a good example of how liberal ideas might turn into neoliberal practice. Another disturbing factor is increasing inequality among public schools, especially schools with a high concentration of immigrants due to flight of middle-class families. This results in distribution of children along fused socioeconomic as well as ethnic lines.

Denmark's liberal Grundtvigian tradition of "education for life" used to be the positive side of the liberal state and the source of much national pride in the Danish school system. Grundtvig (1783–1872), was a Danish author and priest who has become the icon of the Danish school system. He was one of the founders of the "Folkhøjskole"(people's high school), where everyone regardless of class could spend some months to learn new ideas as well as national history. The Højskole has prided itself ever since on being a school for life knowledge, and there are no exams—the exact opposite of the present testing system.

Despite neoliberalism's very conspicuous turn from this tradition, only a few educational researchers and intellectuals in Denmark have generated a strong critique. A promising development has been formation in 2006 of oppositional groups of Danish educators, Concrit (www.concrit.org), which intends to discuss neoliberal restructuring of education in a European context. Another oppositional organization, Sophia (www.sophia-tt.org), has what some consider a more parochial national focus. I think both organizations are still fragile and need to be connected to international movements.

Finally, I can point to a more mainstream reaction (awakening) among educational researchers from *The Scandinavian Journal of Educational Research* (2002, 2006), which has published two special issues that take up the idea that there is indeed a distinctive Nordic school system; it is under attack, but it remains viable, and we must defend it.

References

Carlgren, I. 1994. Curriculum as social compromise or accident? Some reflections on a curriculum making process. In *New policy texts for education*, ed. D. Kallos and S. Lindblad, 1–34. Sweden and UK, Pedagogiska rapporter 42. Umeå: Umeå University, Sweden.

————, ed. 2002. Special issue, *Scandinavian Journal of Educational Research* 46, no. 3.

Frímannsson, Gudmundur Heidar, ed. 2006. Is there a Nordic school model? Special issue, *Scandinavian Journal of Educational Research* 50, no.3.

Simula, Hannu, Ingolfur Asgeir Johannesson, and Sverker Lindblad. n.d. *Changing education governance in Nordic welfare states: Finland, Iceland and Sweden as cases of an international restructuring movement.* n.p.: Taylor and Francis Group.

CHAPTER 6

Higher and Tertiary Education in the West Indies Ensnared by GATS

Marguerite Cummins Williams

In researching this chapter about the effects of GATS (General Agreement on Trade in Services), I have been intrigued by the information I have found, which shows so clearly how developing countries without advisory strength, like Barbados and the neighboring island of Jamaica, can easily be entangled in an expensive coil. At the time of writing, I still have not found out if the Jamaican government has been successful in getting some relief from the commitments it has made to GATS, which may have devastating effects on its vulnerable educational system.

When Jamaica joined the World Trade Organization (WTO) in March 1995, it had five years to enter into negotiations under GATS to achieve "progressive liberalization" of its services by reducing or eliminating "barriers" to market access. Not joining in GATS would have meant risking exclusion from equal access to markets, which could affect the country's exports and therefore its vulnerable economy. Yet once a country like Jamaica has committed itself to opening up its services under GATS, retreat is well-nigh impossible without incurring liability to pay possibly substantial compensation to affected countries or perhaps making a deal in another area as reparation. The hearings on these issues are decidedly nondemocratic since they take place in secret and there is no right of appeal.

The importance of Jamaica's experience is that it fell into the trap of signing up its education services under GATS. The WTO and some of its members talk about an "education market"—a global market opportunity. This is in stark contrast to the gut feeling of most educators and trade unionists and indeed the general public that education is a public good that should not be traded as a commodity, particularly if this interferes with the responsibility of states to provide equitable education for their citizens. Of the 149 member states in the WTO only forty-four, including three in the Caribbean—Haiti, Trinidad & Tobago and Jamaica—have made commitments to trade in education services. Among those making unconditional commitments to higher education are Congo, Sierra Leone, Lesotho, and Jamaica. The major developed countries have made no such unlimited commitments on education, and some commentators have suggested that where developing countries have done so, this has been partly as a result of an incomplete understanding of the implications and in the hope of receiving much-needed assistance. In some cases, "incentives" such as the possibility of prestigious institutions setting up Overseas Campuses (OVC's) in the territories may be dangled before the negotiators.

The strongest opposition to what can only be called this commodification of education has come from the North American and European education unions and associations and from the global union, Education International. Governments like Jamaica who have now committed themselves might have acted otherwise had there been more consultation and openness about decision making with their unions. UNESCO itself has advised its member institutions to refrain from making commitments in higher education and if already committed, to go no further. The Caribbean Regional Negotiating Machinery has also warned that "States that have not made any commitments in their tertiary education sector must proceed cautiously and only make commitments that will serve their best interests. The complexity of the trade negotiations warrants the need for technical assistance on behalf of the developing countries." This body, which was set up to advise Caribbean countries on WTO negotiations, warns particularly about the rapacity of the United States: "As the leading provider of services in tertiary level education (TLE), the US has sought full commitments for market access for higher education and training services, for adult education and for 'other' education.'" It further warns that states belonging to the Caribbean Community and Common Market (CARICOM) "would be well advised not to accede to this request since developed countries are much more competitive in this sector and the Caribbean domestic providers are not ready yet for an open, market driven environment."

Jamaica's commitment to education is typical for the Caribbean. The budget expenditure in 2000 on education was about 15 percent of GDP, with 19 percent of that for higher education. Students' fees used to be totally covered by the government. Now students also pay fees, 10 percent at nondegree level and about 18 percent at degree level for the Jamaica University of Technology and the University of the West Indies. The government is the single largest provider of higher education. In order to upgrade its teaching staff and facilities it has been projected that Jamaica needs to spend an additional JA\$219 billion over the next ten years. Education analysts in Jamaica say that this is an extremely difficult target. And yet due to its commitments under GATS, the Jamaican government could be required to fund a potentially large number of foreign suppliers who would provide education to Jamaicans. Any financial support by government for students or institutions may be considered as subsidies with respect to GATS rules. The issue is further complicated by the way in which GATS documents appear to use the words "higher education" and "tertiary education" interchangeably. For Jamaica, it is at the tertiary level that ill-advised commitment has the most costly and serious implications. There is a tradition in the Caribbean of allowing foreign providers into the tertiary education market on a case-by-case basis. The criteria applied involve examining each to see how it fits into government education policies and development strategies. There are many examples of arrangements such as twinning, partnerships, and branch institutions. But if any part of an education institution functions on a profit-making basis, it automatically runs afoul of GATS rules: it is not considered as totally financed by the government but in competition with private providers—for example, Florida International University, which is foreign owned and operated. Worryingly, too, the government provides the operating costs of several high schools owned by denominations and trusts, so secondary schools too could come under GATS rules.

The money will start to drain from the Jamaican education budget if the government is forced to respond to demands for financial support such as boarding grants and loans from students attending foreign private colleges and universities on the same basis as from students attending local public institutions. So a student at Florida International would expect the same level of financial support as one attending the publicly owned and funded University of the West Indies. Thus from an already stretched education budget, the Jamaican government would be subsidizing private institutions, many of them based in some of the wealthiest countries in the world.

The commitments made by Jamaica when it joined the WTO were not particularly at odds with the then current situation. After all, there were at that time over 150 institutions in the Caribbean, of which 60 percent were public and 30 percent private; the remaining 10 percent existed with some government support. This is understandable given the need to increase the number of young people attending university, which is vital for the development of the region. What was not appreciated, however, were the long-term implications of the commitments made under the GATS system. While a vulnerable developing education system like the Jamaican one is now caught in the net, a spokesperson for the presumably better advised Australian Department of Foreign Trade commented, "Australian commitments were structured so that we have the ability to discriminate between foreign and domestic private institutions (e.g., in relation to subsidies) should this ever be an issue."

Another vulnerable area would be the Caribbean Examinations Council, which provides testing services throughout the region. Its viability is dependent on its market, and so they are vulnerable to competition in particular from distance education and testing services in the United States, especially since they are backed by the attraction of the U.S. colleges whose requirements they have years of experience of satisfying. Another fear is that the GATS commitments enable Jamaica to be used as an insertion point for foreign educational materials to be allowed into the region.

Once the foreign private providers are ensconced and publicly subsidized in Jamaica, the concern from the local public colleges is that the foreigners will provide inferior and irrelevant programs because the profit motive is paramount. There is a perception that foreign providers focus on areas where the necessary facilities and staffing are minimal and leave local providers to finance the more expensive and less immediately popular areas. For example, setting up arrangements for short business or computer courses is cheaper and easier than teaching laboratory sciences.

There are exciting plans in the Caribbean for cooperation between the publicly funded universities in the Caribbean, for instance the Caribbean Knowledge and Learning Network, which will provide the infrastructure for e-learning. CARICOM is committed to funding and supporting the development of the University of the West Indies and national tertiary education colleges, since these have an intrinsic commitment to local societal interests and structure. However if the biggest island in the Caribbean, Jamaica, is burdened by the responsibility of providing the costliest programs against competition from private, for-profit institutions with comparatively vast resources and budgets, its task will be almost impossible. The foreign private sector will be able to attract quality staff with higher wages and better conditions of service, while the local public institutions—which, as a result of GATS, will be progressively more strapped for cash—will suffer from understaffing and a shortage of resources, causing a spiral of decline so that education, and consequently society, will be even more stratified.

It is quite clear from Jamaica's experience that negotiating separately, especially against the large, well-endowed and experienced negotiation machines of the developed countries, is not a winning option. Under GATS, market forces, not social and human considerations, will be the guiding criteria. Commitments made under GATS involve a country in a web of complex rules and punitive measures even when a government is trying to ensure that its citizens have equitable access to quality education as a necessary prerequisite and partner to development. The Jamaican experience is a cautionary tale for developing countries.

CHAPTER 7

The Education World Is Not Flat
Neoliberalism's Global Project and Teacher Unions' Transnational Resistance

Larry Kuehn

Thomas Friedman uses the phrase "the world is flat" as his book title and the metaphor for his explanation of the impact and inevitability of globalization in homogenizing societies. Yet Friedman's analysis does not really reflect reality for most people, those beyond the representatives of global elites that he interviewed and dined with in researching *The World is Flat* (2006). It is certainly true that global brands and global media reach even into seemingly isolated villages almost anywhere. What Friedman omits is that while commodities, including cultural commodities, travel across borders, most people do not and cannot. They still live in communities. Those communities need to be able to maintain education systems that reflect their needs for social cohesion and democratic development and not just be dependent on global markets, cultural "products," and the "anonymous socialization" of the Web 2.0 social networking tools.

However, economic and political forces pushing to "flatten" the world of education are real and take many forms. Education policies are increasingly aimed at achieving global economic competitiveness, and neoliberal ideologues promote privatization as the answer to most problems. Commercialization and corporate intervention encourage the adoption of business methods of management in education. International tests are used as comparative measures, propelled by demands of the international monetary institutions to promote uniform policies. Politicians and education bureaucrats import policies used elsewhere. Publishers of education tests seek to open new markets in different countries. And as I explain in this essay, trade negotiations and agreements treat education as a tradable commodity.

Vignette 7.1

At the end of 2006, 144 teacher union leaders in Pakistan were sacked and the unions banned. A subsequent protest meeting called a general strike to oppose the ban and privatization of schools. From the report "Trade Union Rights Campaign Pakistan (TURCP) in Sindh rejects ban on teachers' organisations," turc-p, http://www.turcp.org/articles/2006/10/23sindh.html (accessed August 12, 2007):

> Trades Union Rights Committee of Pakistan national organizer Khalid Bhatti said the government banned the teachers' organizations to start the privatization of schools and colleges. This ban is the first step in that direction. This government wants to privatize all the main educational institutions in the province which are providing very cheap education to poor working class students. Privatization will make it impossible for working class youth to study in these institutions. The government wants to break the teachers' power to make it sure that there should not be any resistance or opposition to the privatization policy.

Each of these factors plays a role in the pushing for a harmonized global pattern for education. Working together they create significant pressure toward a flattening of education systems so that they look and act more and more alike across national boundaries, even as the communities they are supposed to serve remain quite different in their needs and resources. Although trade agreements are only one of these elements causing a homogenization of education, the transnational resistance to globalization in education has primarily been focused on trade agreements. Trade negotiations provide a face and a place to focus that resistance, primarily taken up by teacher unions, along with their allies who share the sense of importance of public education as a central institution of democracy and equity.

I first explore how trade agreements relate to the rest of the neoliberal program, explaining the context for pressures to include education in trade agreements and the probable effects of doing so. Then I look at strategies, both potential and actual, for teacher unions and some transnational coalitions that resist education's inclusion in the trade agreements.

The Context for Trade Agreements:
Education's Altered Purpose, Funding, and Organization

Central features of globalization are ideology, economic power, and technology—neoliberal ideology, corporate global capital, and information and communications technology. Susan Robertson explains in her contribution to this volume how neoliberal ideology asserts the market's central role in the allocation and distribution of goods and services, pushing the state—that is, a local or national government—to a marginal position in the provision of products, services, and social equity (Spring 2004).

As the state recedes and markets globalize, large corporations have the best chance of operating successfully, and policies favorable to corporate interests are more widely adopted. The decline of the role of the state has significant implications for education, which has been one of the central roles of the state for well over a century, both in most of the developed countries and, particularly in the postcolonial period, in most of the less developed countries. That pattern is being reversed as one element of the neoliberal globalization process. The share of education being carried out by the state is declining, and that by private for-profit and non-profit education is increasing. In some cases, the state funds some portion of this private education through mechanisms such as student subsidies or loans for postsecondary students in private institutions, vouchers, and direct state funding for private schools, such as that provided in British Columbia to "independent schools." Privatization in its various forms is one of the significant elements of the impact of globalization on education.

The World Bank plays a significant role in promoting privatization in the less developed countries, operating through two primary mechanisms. One is its research and publishing. The other is its power of policy imposition through conditions on loans to countries (Spring 2004). Studies such as

---

Vignette 7.2

From World Bank Report "Bringing the Little Girls of Pakistan to School," August 2005. http://siteresources.worldbank.org/INTPAKISTAN/Resources/Pak-educ.pdf (accessed July 7, 2007):

> The education policy package the Bank is supporting at the provincial and local level focuses on free tuition and books, upgrading school facilities, contract teacher recruitment to fill vacancies, stipends for girls in backward areas and *public funding of non-government—independent, low-cost private—education*, eventually through grade ten [*emphasis added*].

Education: The World Bank Education Sector Strategy (World Bank 1999) outline the World Bank's directions for education, based on the assumption that a market economy requires a primary focus on the development of "human capital," enhanced individual skills for economic competition. These skills must be upgraded on a constant basis through "lifelong learning," because in globalized markets corporations will move production to where it finds the best employees at the least cost. All of this is best carried out by more flexible service deliverers, primarily private or non-profit groups, rather than the state. If the state continues to maintain responsibility, then the service is decentralized or "municipalized," with limited control by the state. A section of the World Bank Web site is devoted to providing both a rationale and tools for privatization (http://rru.worldbank.org/paperslinks/public-private-education/).

Conditions set by the World Bank for loans to countries in difficulty, "structural adjustment" policies, require reduction of barriers to import of goods and services and a reshaping of production to focus on export industries. Another requirement is reduction of the role of the state, both in regulation of the economy and the provision of services. Since education is the one of the largest areas of state expenditure for most of the less developed countries, reductions of public funding and delivery of education is a common factor in neoliberalism's globalization.

In the early 1980s the World Bank signaled a decisive move away from development as a process of national economic growth to embrace a vision of development as "equal to participation in and integration with the capitalist market" (Munck 2002, 6), incorporation into the global marketplace. As Jones says in his study of the World Bank and financing education, "It is now possible to speak of an international system of influence powerful enough to bind up the educational destinies of the world's peoples" (Jones 1992, xiv).

This binding of destinies is not confined to those who live in the less developed countries. However, the mechanism for this common direction has been "policy borrowing" rather than policy imposition: "The key features of policy borrowing . . . are that it is carried out voluntarily and explicitly, and that its locus of control is national. It involves particular policies that one country seeks to imitate, emulate or copy, bilaterally from one another. It is the product of conscious decision-making and it is initiated by the recipient" (Dale 1999, 9–10)

Policy borrowing is carried out sometimes through politicians looking elsewhere for ideas for dealing with particular issues that have become a matter of public concern, but at other times, bureaucrats within the education structures or academics bring to decision making the ideas and policies adopted elsewhere. As an example, when the Asia-Pacific Economic

Cooperation (APEC) created an education forum of ministers of education from the APEC countries, one of its early activities was a conference on the OECD education indicators, hosted by the U.S. Department of Education, even though most of the APEC countries are not a part of the OECD (Council of Ministers of Education Canada 1995). Mishra (1998) contends that the activities of international organizations "amount to a supranational steering of social policy in a neoliberal direction," with global institutions that "influence social policy largely by means of policy prescription, expert advice and general economic surveillance" (491).

Policy borrowing in education is certainly not a new phenomenon. Ideas on education have been borrowed and built upon at least since classical Greece. But what seems to distinguish the education policies of neoliberal globalization in both the most and less developed countries are two primary elements: commonality about the purpose of education and the reform of structures of educational governance and control. While each of these takes on a shape influenced by local circumstances, at the root they aim to create "appropriately skilled and entrepreneurial citizens and workers able to generate new and added economic values, [that] will enable nations to be responsive to changing conditions within the international marketplace" (Robertson 2000, 187).

While previous policies aimed to prepare students for participation in the economy, what is new is pressure on education systems to reorganize so that they increasingly aim to prepare students for skills and competencies required by workers in a globalizing, information technology-based world (Chan-Tiberghien 2004, Au and Apple 2004). Levin (1997) identifies common themes in globalized education reforms in governance and operation in English-speaking countries as local management, choice and markets, and testing and accountability, but these themes are also global in nature because the World Bank prescriptions "frequently reflect the current school reform proposals in the United States" (Spring 1998), to the point that "charter schools" is a term used by Spanish speakers in Argentina.

The elements Levin demarcates all flow from the ideology of neoliberalism. The apparent role of the state is reduced by decentralization. The responsibility for education is given to municipal levels of government or to parent committees for individual schools based on the claim that doing so makes the schools more responsive and responsible to communities. The state often then reduces the financial support that it provides, leaving either the municipal government or the parents to pick up financing of the schools. With this decentralization, it is possible to open up to "choice," where different schools have different characteristics and a parent can opt to choose a particular type of school. The intention is to produce a market-like situation, even if all of the schools receive public funding. In neoliberal theory, this kind of market choice will produce higher quality because of the competition to attract students to a particular school. Schools will improve because they change to attract more students.

Testing and accountability are also supposed to motivate improvement. The data produced is to provide parents with information to make effective choices and to provide the state with tools to direct the schools in what knowledge and skills are to be taught, as well as to provide data to use to direct changes to be made in specific schools. The testing and accountability system provides a way of the state "steering from a distance." The state reduces the degree to which it is a direct provider and financier of educational service, at the same time having more effective tools to direct the intended outcomes of the educational process. The testing regime has been expanded to go beyond the comparison of school to school to produce international comparisons, a policy template that fits with configuring education as a factor in global competitiveness and more closely aligns expectations and policies in relationship to education. This process is apparent in the Organization for Economic Cooperation and Development (OECD) creation of an international indicators system to provide comparative data on "inputs" and "outcomes." The OECD developed its own comparative testing program called PISA to have data to shape education policy. The OECD Education Indicators Project is primarily interested in education as an economic factor, as was intended when the Reagan administration took away from UNESCO the resources to build an international comparative system and gave it to the OECD (Kuehn 2004).

Space does not permit me to explore the powerful critiques of this pattern of decentralization, marketization and testing, but many analysts, like Levin (1997) and Hargreaves (2003), have pointed to the contradictory purposes of the reforms, which have a weak connection to teaching and learning—and outcomes.

Globalization of education has produced common themes across countries, whether by "policy borrowing" or "policy imposition," creating the basis for challenges to these ideas to be made on a transnational basis as well by those seeking a counter to neoliberal forms of globalization. I suggest that while the challenge must be transnational, it must at the same time respect local and indigenous formulations and democratic control and practice in education.

Trade Agreements as Wedge and Ratchet

Although trade agreements have not been the primary driver of globalization of education, as they have developed in the last two decades they do play two

---

Vignette 7.3

From Sudha Ramachandran, "Literacy Beats Out Education in India," *Asia Times*, April 24, 2004, http://www.atimes.com/atimes/South_Asia/FD24Df04.html (accessed July 7, 2007)

> Literacy skill is all that the masses need, argue the market forces, so that they can read the product labels and advertisements. Its somewhat evolved form would be adequate for factory workers to read production instructions and to use even the Internet. Critical thinking, creativity, scientific temper, analytical abilities, sense of history or philosophy, aesthetic appreciation and other such educational attributes need to be reserved for the privileged few—this is the implication of the literacy paradigm and the market forces.

---

important roles in the globalization process. The agreements function like a wedge to open up more and more aspects of society and economy to the trade regime, including education. Further, they serve as a ratchet that allows only more privatization, never allowing the return of what has been privatized back into the public sector.

Key to the wedge aspect is acceptance of an international legal framework that commits all parties to agree not only to the existing rules, but to ongoing negotiations in the direction of trade liberalization, putting more and more limits to democratically made decisions that might go in a different direction. This one-way path is set out by the Services Council of the WTO in the General Agreement on Trade in Services (GATS) negotiations when it says that "members shall aim to achieve progressively higher levels of liberalization with no a priori exclusion of any service sector or mode of supply" (Khor 1994).

This international system advantages trade and economic values over social concerns, both in political discourse and allocation of power. While many international agreements exist, some can be enforced, but others only require voluntary adherence. For example, the World Trade Organization (WTO) rules protecting "free trade" are enforceable, but the conventions of the International Labor Organization (ILO) protecting worker rights are not, even though ILO conventions, like WTO rules, are international treaties agreed to by nations. Similarly, the trade rules of NAFTA are enforceable, but the labor and environment side agreements are not. Both the ILO and the NAFTA side agreements operate by complaints, investigations, and reports, but with no compulsion for countries to abide by the recommendations from the reports. In contrast, breaking the WTO and NAFTA trade rules brings economic consequences for those nations and industries that lack the political leverage to contest the rules with success, as occurred when the powerful U.S. lumber production lobby forced Canada to agree to quotas on lumber competition from Canada.

The second key role of trade agreements in neoliberal globalization is to freeze democratic control of the economy through a ratchet effect. A country that has agreed to the rules of the WTO and NAFTA can never change its policy to bring in new state regulation or return services that have been privatized to the public sector without a great penalty. As Margueritte Williams' discussion of effects of trade agreements on the University of West Indies in this volume indicates, education officials and trade officials negotiating agreements often have not been aware of the impact on education until after negotiations have been completed. Yet, the explicit commitments in the agreements to move toward further liberalization of rules, produce an erosion of democratic control. It limits the ability of a government to change directions, even if the populace votes for the change and wants to take services outside the trade regime—thus these agreements are a freezing of democracy.

Vignette 7.4

From Tony Blair's speech, "Doctrine of the International Community," given at the Economic Club, Chicago April 24, 1999. http://www .number-10.gov.uk/output/Page1297.asp (accessed July 7, 2007):

There is a new economic role for government. We don't believe in laissez-faire. But the role is not picking winners, heavy handed intervention, old style corporatism, but: education, skills, technology, small business entrepreneurship. Of these, education is recognized now as much for its economic as its social necessity.

Trade Negotiations Provide a Place and a Face to Globalization

Though trade agreements are not the primary tool for globalization of education, they have become an important focus in organizing opposition to neoliberal policies in education. Why?

A key factor has been that publication of the terms being negotiated in trade agreements makes visible and concrete ideological elements of globalization and their potential impact. As an example, the "requests" from the United States Trade Representative (USTR 2000) in the GATS negotiations that educational tests be considered as a tradable service covered under the GATS illuminate that the creation and marking of educational tests is being considered an economic function rather than a primarily educational issue. Tests are on the negotiating table not because the tests have proven educational value but rather because they are a potentially profitable business for corporations. A country's desire to choose people from its own education system to create and mark tests places a limit on testing services as profitable economic activity, contradicting the view of testing put forth by trade negotiators.

Looking at the text of what has been or is being negotiated is a concrete way of understanding the ideological direction behind it. Examining the text and the process of negotiating trade agreements is a strategic approach to get attention to the nature of the changes taking place. Further, because trade agreements go through a negotiation and ratification process that has some public element, they can become a focus of action.

There is no concrete place or face to the ideology of neoliberalism. On the other hand, there are places and faces to a trade negotiation—Seattle, Doha, Quebec City, Genoa, Cancun, Miami, and on and on. Trade negotiators would have been happy to continue their work in the back rooms without any public attention. However, a number of civil society groups have brought the attention to these negotiations and their impact: unions and international labor bodies, environmentalists, progressive think tanks like the Canadian Centre for Policy Alternatives and the Third Word Network, and activist groups like the Council of Canadians. The resistance of social justice movements to other corrosive economic and political effects of trade agreements provides a "home" for education activists and teacher unions who see the injuries being visited on public education by neoliberal policies.

Popular opposition to elements of the trade agreements has a history as long as the agreements themselves. However, only gradually has the harm to public education been realized. The first trade treaty to include services, including education, was the Canada-United States Free Trade Agreement of 1988. This agreement was opposed in Canada by a coalition of labor and other progressive organizations, as well as the Liberal and New Democrat parties, in the election of 1988. The opposition focused on the nature of the relationship of Canada with the United States and the groundbreaking element of including trade in services was little recognized. While the Canada-United States FTA was a major issue in Canada, it was little known in the United States.

NAFTA, on the other hand, had a high profile in all three countries—Mexico, Canada, and the United States. Opposition in Mexico came primarily from labor unions independent of the control of the PRI, the dominant political party in Mexico. This included the "democratic current" opposition within SNTE (Sindicato Nacional de Trabajadores de la Educacion). This opposition, described elsewhere in this book, had the majority in some states in the south of Mexico and among the elementary teachers in Mexico City, as well as a presence in other states.

In the United States, both the National Education Association (NEA) and the American Federation of Teachers (AFT) opposed NAFTA based on the general labor concern that it would lead to the loss of jobs in the United States—not on an analysis that it would have a particular negative impact on public education. The same labor and progressive organizations in Canada that opposed the Canada-United States Free Trade Agreement also opposed NAFTA. However, by the time NAFTA was being negotiated in the early 1990s, Canadian activists had an awareness of the possible implications of including services within trade agreements. One of the elements specific to education that was initially identified as significant dealt with the certification of professionals. Under the NAFTA definition, public school teachers clearly fit the definition of professionals covered by the proposed agreement.

The model for transnational recognition of common professional qualifications in NAFTA was that of accountants. This is not surprising, since businesses that operate in a range of different countries would find it an advantage to have accounting and auditing carried out following rules and standards that provide comparability. However, the further one gets away from business into social and cultural areas, the less sense it makes to insist on common recognition of the professional qualifications. At least that was the perspective put forward by the British Columbia Teachers' Federation (BCTF) in its response to the proposals for NAFTA.

Professional certification requirements were an area of potential impact of NAFTA discussed at some length in the BCTF brief to the Canadian Parliamentary Committee on the North American Free Trade Agreement in December 1992. Section A.1 of Annex 1210 of NAFTA states, "This annex applies to measures adopted or maintained by a Party relating to the licensing and certification of professional service providers." The BCTF described some differences in the certification practices in the United States, Canada, and Mexico and put forward its concern:

> The point here is not to argue that one is the better than the other, but that they are significantly different. Those differences reflect the fact that certification of teachers is imbedded within complex cultural practices that make up education. Because of that, they are not easily changed, nor should they be.
>
> But because NAFTA is based on a view that sees education and all other services as primarily economic exchanges, it demands negotiation to make the practices in each of the countries more similar to each other.
>
> Specifically, Annex 1210 calls for "development of mutually acceptable professional standards and criteria." The areas defined for common standards

include "conduct and ethics, professional development and re-certification and scope of practice." It is in these areas, as indicated above, that there are differences between what generally applies in Canada and the US. (British Columbia Teachers' Federation 1992, 7)

Concern about certification was fueled by a paper from the University of Southern California and the Educational Testing Service that included a proposal for

transnational educational certification, e.g. teacher certification, language proficiency certification, technical certification, and other forms of professional certification and testing standards.

A trilateral commission should consider how the three North American countries will determine fitness to work or provide services. The establishment of a common set of education standards could be formally enforced through a certification system acceptable to all three partners. (University of Southern California and Educational Testing Service 1992, 5)

Though the idea has been in NAFTA's presence for more than a decade, it has not been pursued seriously. The NAFTA commission working on common certification has mostly focused on areas directly related to business, such as engineering. Both the United States and Canada have degrees of state and provincial authority over education, creating barriers to common certification recognition. However, Canada is moving toward common recognition within the country, the initial step necessary to create common certification in the longer term. In the United States, a multiple-choice test, created through a grant from the U.S. Department of Education, is used to certify teachers in five states. To the degree that education becomes homogenized through various other aspects of globalization, issues of common certification are more likely to arise.

Growing International Resistance: Creation of the Tri-National Coalition in Defense of Public Education

The transnational nature of trade agreements seemed to demand that effective responses would require a transnational basis. While international teacher organizations and unions were involved in defending public education in respective members' countries, in the early 1990s they had not developed a consciousness and program on globalization related to trade agreements.

The first of the international teacher union coalitions that grew out of trade agreements was the Tri-National Coalition in Defense of Public

Education. In January 1993, the Labor Center at Evergreen State College in Olympia, Washington, brought together 200 educator-activists from Mexico, the United States, and Canada. Concerned about attacks on public education reflected in trade agreements, the participants drew up a strategy for protecting and promoting public education in North America in the face of further economic integration along neoliberal lines. The strategy involved forging a long-term coalition of teacher unions at the national, state, and local levels. A fledging organization was built through a number of meetings coordinated by the then Labor Center Director at Evergreen, Dan Leahy, who took a sabbatical in Mexico. Following the first of those meetings, important Mexican and Canadian teacher unions formally joined the fledgling coalition, including the unions representing instructors at the Universidad Nacional Autónoma de Mexico (UNAM) and the Universidad Autonoma Metropolitana (UAM) in Mexico City, the Michoacan section of SNTE, the Canadian Teachers Federation, and provincial unions in Quebec, British Columbia, and Ontario. The NEA sent participants to an initial meeting, held in Zacatecas, Mexico, in 1994. However, the NEA declined to participate in the coalition, with the leadership indicating that the effects of neoliberal trade policies on education were not clear. As with other U.S. unions, NEA's opposition to NAFTA had been based on the more general concerns about jobs and industry and not on analysis of its potential impact on education. The AFT did not respond to invitations to participate.

Broader political considerations also affected NEA's participation. Teachers' organizations from Mexico joining the coalition did not have a majority in SNTE, and SNTE supported NAFTA and the neoliberal policies being followed by the Mexican government. Complicating the situation was the fact that both the NEA and the SNTE, which are national unions, belonged to Education International (EI), the international confederation of teacher unions, with some 30 million teachers in its member organizations.

Despite lack of participation by the U.S. teacher unions, the Tri-National Coalition in Defense of Public Education developed as an organization with a range of activities in opposition to neoliberal policies. The Mexican participants created a structure that brought together unions representing university faculty from several universities, along with teacher unionists participating in the CNTE (Coordinadora Nacional de Trabajadores de la Educacion), the opposition grouping within the SNTE. This Mexican Section of the Tri-National Coalition has a coordinating committee and holds meetings where representatives of all the participating groups carry out analysis of issues and develop reports. The Mexican Section has been active in organizing and sending participants to the seven Tri-National Conferences in

Defense of Public Education held between 1994 and 2006, five in Mexico and two in Canada.

In addition to the international participation, the Mexican Section has carried out campaigns within Mexico in opposition to specific neoliberal policies that have followed from the integration that NAFTA initiated. For several years the Mexican Section has campaigned in opposition to the *examen unico*, a standardized test being administered as a requirement for postsecondary education placement. The test is administered by CENEVAL, a nongovernment agency modeled after the Education Testing Service. This agency was created after NAFTA as a part of the patterning after the United States in the homogenizing process. The Mexican Section has also been active in opposing the restructuring of the public pension plans proposed to reduce the contribution of the state to the pensions of public employees.

The seventh Tri-National Conference was held in Oaxaca, Mexico, in March of 2006, just six weeks before the beginning of the teacher strike in that state that exploded into a social struggle engaging social groups well beyond just teachers. The Tri-National Coalition played a role in developing international solidarity among unions outside Mexico in support of the teachers in Oaxaca in the strike and occupation of the center of Oaxaca City that went on for nearly six months. That conference in Oaxaca also marked the first significant participation of U.S. unions in the Tri-National, with delegates from the AFT local at the City University of New York and from the United Teachers of Los Angeles, a joint AFT/NEA local.

Canadian provincial teacher unions have played a major role in supporting and participating in the Tri-National Coalition throughout its existence. The BCTF international solidarity program has provided funding, and the union has played a coordinating role on an ongoing basis. The national umbrella for English Canadian teacher unions, the Canadian Teachers' Federation, has participated at most of the conferences and some university and college faculty unions and the Canadian Association of University Teachers (CAUT) participated in conferences, as well.

Building Enduring International Coalitions

The nearly fifteen–year life of the Tri-National Coalition illuminates the challenges in constructing and maintaining international coalitions. One of the key conditions for a coalition is a common understanding of the context and the consequent need for a coalition. NAFTA was seen as a concrete manifestation, as well as a symbol, of the neoliberal integration of economies and thus a threat to public education by at least some union activists in Mexico and Canada. Those who saw this as a significant need organized around the issues, even if they represented only a part of the teacher unions.

The first stage of building the coalition was information. The 1993 conference at Evergreen State College was key to this. It was the first time that teacher union activists from the three countries had been together to share information about the realities of education in their countries and how neoliberal policies might affect these. Out of that conference, reports and further analysis was developed, again particularly in Canada and Mexico. Both of these countries are in very unequal relationships with the United States, meaning that harmonization is likely to require the most changes from those two countries, thus creating more concerns among unions representing teachers in those two countries.

Because coalitions are not usually formal organizations, they often appear around an issue then disappear. If there is no formal affiliation and organizational structure, they are likely not to last. The Tri-National has confounded this primarily for two reasons: continuity of leadership and resources. A person in each of the three countries was prepared to undertake maintaining the coalition as a project and each has continued to do so over the fifteen years. The resources for the activities have come from a number of unions, but the continued support of the BCTF International Solidarity Fund has provided a consistent financial base for activities. These leadership commitments and resources have been key.

While the Tri-National Coalition began in response to NAFTA, over time the focus has expanded beyond the impact of trade agreements and the three countries.

The Coalition Expands to the Hemisphere of the Americas

Just as NAFTA was the impetus for the formation of the Tri-National Coalition, so the Free Trade Area of the Americas (FTAA) prompted formation of a similar but broader coalition that encompasses the Americas, North, Central, and South.

The U.S. initiative to expand NAFTA to cover the Americas in the form of the FTAA was intended to be developed through a series of summit meetings, bringing together the leaders of all the countries in the Americas (except Cuba). The second summit meeting, in Santiago, Chile, in 1998, adopted a program of action on education that were supposedly commitments of all the governments, in addition to continuing the negotiating of a trade agreement.

An alternative People's Summit was held at the same time, organized by a Hemispheric Social Alliance and a regional labor organization for the Americas.

Despite education being a major topic at the official summit, the education component of the People's Summit did not provide an alternative program that challenged the neoliberal ideas about the role of education that were reflected in the official program adopted by the governmental leaders. In response to this gap in presenting an alternative to neoliberal education policies, CoDevelopment Canada and the BC Teachers' Federation, proposed a conference to develop an alternative program. CoDevelopment Canada is an international NGO that works with the BCTF and other unions on international solidarity projects in the Americas. The two organizations invited participants from regions around Latin America to meet in conjunction with the 1998 Tri-National Conference in Defense of Public Education to explore forming a broader coalition to focus on the FTAA and the education commitments of governments made at the Santiago Summit.

This group of a half-dozen participants organized a conference in Quito, Ecuador in 1999 to explore education issues related to the negotiation of an FTAA and the Santiago commitments on education. This IDEA Conference (Initiative for Democratic Education in the Americas) created the IDEA Network (Red-SEPA in its Spanish acronym), which included teacher unions but also student and community organizations committed to public education. The IDEA Network participated in alternative conferences at the next two Summits of the Americas in Quebec City in 2001 and Mar del Plata, Argentina, in 2005, as well as at the World Social Forum and the World Education Forum.

As with the Tri-National experience, the focus of the IDEA Network has gone well beyond trade agreements to include research projects and targeted meetings on a range of issues related to neoliberal policies, including standardized testing, decentralization of funding, and gender issues.

A paper adopted at the initial IDEA conference identified several elements of strategy:

- Defend public education at the local and national levels with a strategic consciousness of the global context. Inform and mobilize teachers to take part in this defense.
- Counter neoliberal ideology with an alternative program for public education nationally and internationally.
- Conduct research and analysis and share it with other organizations throughout the Americas.
- Build communication links among organizations with conferences and with communication using the Internet.

- Work in international and regional teacher and labor organizations (e.g., Education International) to develop common understanding and common strategies.
- Take part in international campaigns aimed at achieving social rights, including the right to an education and the right for workers to form organizations that provide protection.
- Constantly challenge the "cult of the inevitable"—the claim that there is no alternative to neoliberal policies.

With varying degrees of success, these have been elements of the work of the IDEA Network. The IDEA Network has produced publications and posters as elements of campaigns, organized support actions around the Americas to support teacher unions in conflict with governments, held conferences on specific issues, and created a research network of union researchers, and it is developing an online journal examining issues related to trade agreements and neoliberal education policies.

Opposing Education in the General Agreement on Trade in Services (GATS)

As preparations were being made for the Quito conference, advocacy organizations working on trade issues identified the proposed expansion of the GATS as one of the elements that was on the agenda of the WTO. This information became part of the discussion at the IDEA conference and a central issue in campaigns that would be carried out by the IDEA Network.

The GATS had actually come into existence with the formation of the World Trade Organization in 1995. Because it is a part of the WTO, it covers all countries that belong to the organization. However, the commitments made by most countries on education were very limited, and its existence had not been noticed even by activists involved in campaigning on trade issues. It came to wider notice primarily because it was one of the central topics to be on the agenda for the WTO meeting to be held in Seattle in December of 1999.

Negotiations on the GATS have been ongoing, but fortunately with little achieved by those trying to expand its coverage. The lack of success with the GATS has been largely a result of the failure of other aspects of the WTO negotiations, particularly in agriculture. The so-called Doha Development Round was supposed to provide gains for the less developed countries. Expansion of openings for services in the less developed countries would have been one of the tradeoffs for improved market access in the United States and Europe for products from the the less developed countries.

The service sectors represent a significant and growing aspect of all economies, but particularly those in the most developed nations, which are pressing for all forms of education to be included in the GATS. For the United States, for example, services make up more than 65 percent of Gross Domestic Product and 80 percent of employment (United States Trade Representative 2005). The United States has a huge trade deficit with the rest of the world, but education has consistently produced a trade surplus for the United States, and with a significant portion of its economy made up of services, the export of services is key to reducing trade deficits.

Postsecondary education already reflects a significant export/import market for the United States, Europe, Australia, and, to a lesser degree, Canada. The WTO defines four types of trade in services as the supply of a service:

1. Cross-border supply: from one territory to another, such as education over the Internet
2. Consumption abroad: students studying abroad
3. Commercial presence: foreign direct investment, such as universities with international operations or companies selling tests
4. Presence of natural persons: providing service on a temporary basis in a foreign country

The WTO also makes clear that "the sector includes primary, secondary, postsecondary and adult education services, as well as specialized training such as for sports."

EI has coordinated a campaign in opposition to the inclusion of education in the GATS and has worked along with another global confederation of unions, Public Services International, in the campaign against public services as a whole being a part of the GATS. Together these union internationals make up the largest share of the global membership in trade unions. EI's campaign has included resolutions at its tri-annual congress, monitoring of the negotiations, and taking teams of union representatives to Geneva to meet and lobby national representatives in negotiations. In addition, EI publishes "TradEducation News," an online newsletter that follows the state of negotiations and opposition to inclusion of education in the GATS and has on its web site a GATS Information Kit that details the positions of all countries in the GATS negotiations related to education (www.ei-ie.org/gats/en).

What Lies Ahead for Trade in Education and Transnational Opposition?

The Summit of the Americas in Mar del Plata, Argentina, in 2005 was a disaster for those pushing for a Free Trade Area of the Americas. The countries could not even agree on a communiqué saying that they did not reach an agreement. Despite periodic rumors of new negotiations breaking out at the WTO, new agreements on the GATS seem unlikely although always possible. However, two new directions are apparent. One is that of an alternative to a United States–dominated trade bloc in the Americas. When Venezuela joined the MERCOSUR (Southern Cone trade bloc) after the FTAA failure, MERCOSUR grew to encompass the major economies of Argentina and Brazil, along with Uruguay and Paraguay, with Chile as a sometime associate. The teacher unions in two of those countries, Brazil and Argentina, have agreements made by their national governments that they will not include education in trade agreements.

The other is the negotiation of bi-lateral agreements that build a new trading regime one country or region at a time. These bi-lateral agreements allow nations to bypass obstacles to global agreements. Both the United States and Canada, separately, have reached a number of free trade agreements on a bi-lateral basis, for example with Chile. The CAFTA, the Central America Free Trade Agreement with the United States includes most of the countries of the region, along with the Dominican Republic. Agreements with the United States were reached with the Andean countries, with the recent elections returning regimes in support of the agreement in Colombia and Peru but opposed in Ecuador. Both the United States and Canada are negotiating FTAs with South Korea and other countries in Asia.

As Education International points out in TradEducation News, the education commitments in these bi-lateral agreements often are more extensive than those existing in the GATS or proposed for expansion of the GATS. Teacher unions have been in the lead in opposing many of these agreements. In Guatemala, for example, the union and its leadership have been targeted by the government after playing a central role in the demonstrations opposing the country joining CAFTA. A Costa Rican teacher union had a "Vote no!" campaign in a national referendum on whether to ratify CAFTA.

The building of international coalitions to challenge education being considered a tradable commodity covered by trade agreements depended to a significant degree on the multinational nature of the negotiation of those agreements. It was the formal public meetings that provided the face and the place for organizing resistance to trade agreements, but also to the many other aspects of neoliberal education policies.

With the failure of the large-scale, multicountry form of negotiation to reach new agreements, these meetings as a way of focusing opposition becomes more problematic. The challenge for teacher unions and others seeking to protect public education from destruction through neoliberal policies is to find new points of focus that allow for the deepening of teacher union internationalism in support of an education world that is not flat but reflects democratic control of public education, including concern for indigenous values, social justice, and social diversity.

Vignette 7.5

Extract from the "World Platform of Struggle," adopted by the Charter of the Forum of Education in 2001 at the Third World Education Forum in Porto Alegre in 2004 and by the World Social Forum in Caracas in 2006. "Appeal for the European Week of Action on Education," http://www.ntua.gr/posdep/International/EuroWeek2005/53_Revised_Appeal.doc (accessed July 7, 2007):

1) Education as a public good is a global priority and an inalienable human right which influences people's whole lives.
2) This right is essential for the access to all the other rights. Education is a big aim in the social needs of workers, for the construction of values based on solidarity and for the full deployment and practice of citizenship.
3) The state and public policies in general must ensure that all the necessary means are provided for the realisation and promotion of these rights.
4) It is the duty of the state to guarantee, as part of public service and without discrimination or exclusion related to nationality, religion, ethnicity, sex, social class and sexual choice, the full right to a substantial high-quality public education, at all levels from pre-school to university.

References

Au, W., and Apple, M. 2004. "Interrupting globalization as an educational practice." *Educational Policy* 18 (5): 784–93.

British Columbia Teachers' Federation. 1992. "Presentation to the Parliamentary Committee on the North American Free Trade Agreement." Vancouver, B.C.: B.C. Teachers' Federation.

Chan-Tiberghien, J. 2004. Towards a 'global educational justice' research paradigm: Cognitive justice, decolonizing methodologies and critical pedagogy. *Globalisation, Societies and Education* 2, no. 2: 191–213.

Council of Ministers of Education Canada. 1995. *1993–1995 Annual Report.* Toronto: Council of Ministers of Education Canada.

Dale, R. 1999. "Specifying globalization effects on national policy: a focus on the mechanisms." *Journal of Education Policy*, 14 (1): 1–17.

Friedman, T. 2006. *The world is flat: A brief history of the twenty-first century.* New York: Farrar, Straus and Giroux.

Hargreaves, A. 2003. *Teaching in the knowledge society: Education in the age of insecurity.* New York: Teachers College Press.

Jones, P. 1992. World Bank Financing of Education: Lending, Learning, and Development. *Comparative Education Review* 40 (1): 86–89.

Khor, M. 2004. Preliminary comments on the WTO's Geneva July decision. *Third World Network.*

Kuehn, L. 2004. Leaning between conspiracy and hegemony: OECD, UNESCO and the tower of PISA. In *Passing the test: The false promises of standardized testing*, ed. M. Moll, 57–66. Ottawa: Canadian Centre for Policy Alternatives.

Levin, B. 1997. The lessons of international education reform. *Education Policy* 12 (4): 253–56.

Mishra, R. 1998. Beyond the nation state: Social policy in an age of globalization. *Social Policy & Administration* 32 (5): 481–500.

Munck, R. 2002. *Globalisation and labour: The new "great transformation."* London: Zed Books.

Robertson, S. 2000. *A class act: Changing teachers' work, globalization and the state.* New York: Falmer.

Spring, J. 1998. *Education and the rise of the global economy (Sociocultural, political, and historical studies in education).* Albany: State University of New York Press.

PART III

The Need for Unions to Defend Public Education

CHAPTER 8

Teachers and their Unions
Why Social Class "Counts"

Kathleen A. Murphey

Social class is not a term used willingly or freely by teachers, in or out of teacher unions, in the past or currently. It is not a term used willingly or freely by other working people in the United States. The attainment of middle-class lifestyles by U.S. workers has historically confused worker identity that could arise from the particular work and the working conditions. As industrial workers won the right to organize, their unions and the industrial workers they represented became synonymous with the working class. Labor history has been, up until relatively recently, the history of unionized male workers in the industrial and craft sectors. In the U.S., public sector, unions started organizing in the 1960s. At that time the two teachers' organizations, the American Federation of Teachers (AFT) and the National Education Association (NEA), began to compete in collective bargaining elections to be the "exclusive bargaining agent" for a district's (local education authority) teachers. The AFT had always identified itself as a union, affiliated with the AFL-CIO; the NEA, though competing to be a collective bargaining agent— that is, to represent teachers in contract negotiations with school systems— only relatively recently acknowledged its union status.

Public sector employees, especially teachers, have welcomed unionism, but their identity as workers, as being working class, has always been problematic. They have often seen themselves as professionals, with advanced training, unlike industrial workers. Teachers want the protection of benefits and economic and legal security that their unions offer, but they, as well as their unions, view themselves as professionals, albeit professionals in a union. For some teachers, especially those who come from working-class families, being a professional and a union member are two sides of one coin. For others, such a combined identity has not come easily, and for still others, it remains elusive.

Social class analyses of U.S. education have had periods of vogue: in the 1930s, in the 1960s, 1970s, and, to some extent, the present, when they are newly emerging in university classrooms and in the work of researchers. More often than not, however, insights about social class do not weave their way through the curriculum in teacher education. The more immediate concerns prospective teachers have with mastering methods of teaching usually overshadow discussion of social class and its impact on teaching, teachers, and their work.

The social class lens, however, when it has been applied to education has focused on P-12 urban students, i.e., working class youth, or the working class families and communities from which they come. Teachers in these analyses are (1) sometimes invisible and uninteresting to the researchers, perhaps because the majority of teachers were and are women; (2) seen as pawns in a larger, repressive system; (3) seen as the oppressors themselves, servants in a hierarchical system that upholds middle-class or ruling-class values to the detriment of the working-class students, their working-class parents, or their working-class communities. That teachers are workers themselves and thus need to be included in the analyses is absent from the student, parent, or community focus of these analyses. Teacher unions, the organizational voice for teachers, often seem nonexistent or are portrayed negatively by antiunionists who think unions are not "professional" organizations, or by critical unionists who fault teacher unions for being interested only in economic, not political, issues.

Also, in the field of educational history, teachers as a labor force, up until relatively recently, have had little place. The study of teacher unions has been even more scant. Educational historians viewed teacher union history as belonging in labor history, not educational history. At the same time, the field of labor history has marginalized the history of public sector unions, particularly those of teachers (Shaffer 2002). The consequence is that teachers, as well as the public, are unaware of teachers' labor history.

Roadblocks to Teachers' Social Class Consciousness

Teachers' social class blindness, or unconsciousness, has been perpetuated and protected, usually consciously by both antiunionists and unionists to accomplish their different ends, i.e., to break or make a union. The ideology of professionalism has been used to build a union, with promises that the union is in fact not a union, as well as to destroy a union by confusing, clouding, redirecting, or removing the class question. The unnamed elephant in the room is the hope, or fear, of the power of the educational workforce with a united class consciousness and class solidarity.

Class consciousness has also been confounded by questions about whom the working class includes. Aside from the question about whether teacher unionists are professionals or workers, white men historically have held the jobs in crafts and industry that were first unionized. Working women have had a harder time being considered working class, as they historically were assumed to have taken the class identity of their husbands or fathers. Minority workers, particularly black workers, men and women, were often viewed as outside the class system, their major identity coming from their racial identity. Class and race, as analytical perspectives, clashed violently in New York City in the Ocean Hill-Brownsville dispute in 1968 which pitted New York City's United Federation of Teachers against the Community Control Movement of the black community. Thus, both gender and race have in the past complicated the class question, and they do so in the present (Dolby and Dimitriadis 2004; Perlstein 2005).

Class consciousness has, additionally, been mystified by the ever changing nature of the workplace, relationships at work, working conditions, and the workforce that does the work. In other words, the workplace doesn't stand still; it has evolved historically, and is very visibly evolving now. Thus, the composition of the workforce changes, as does the workplace itself, as do ideologies about work and about the various segments of the workforce doing the work. These ongoing transformations impact class consciousness, posing both challenges and opportunities.

Raising Class Questions Presently and in the Future

What are the future possibilities for class identity and class consciousness of teachers and their unions in a globalizing world? Several important studies cited in Susan Robertson's essay in this volume address this topic. With the political and economic environment, as well as communication technologies, changing so visibly and so radically, all working people, in the United States and around the world, are adjusting to new realities, sometimes exhilarating, sometimes frightening. This has entered teachers' lives most conspicuously with the accountability movement: accountability of teacher education institutions, teachers, educational systems, and P-12 students (Carnoy, Elmore, and Siskin 2003; Furhman and Elmore 2004; Lipman 2004). As teachers and their students face the constant glare of public scrutiny and reforms in the conditions of work, they can no longer believe "this, too, will pass," as they have in the past when confronted with quick fixes to the educational system. Teachers seek leadership in helping them understand, accept, or resist these current reforms. Two things, however, are becoming increasingly clear to teachers: the economic forces that threaten their jobs and public education, as we know them, are part of a reordering of the global economy and the structure of work, and they, the teachers, need more education to understand the depth and breadth of these economic forces that are impacting the world, and especially the world of education.

Teacher organizations have an exciting new opportunity to face the work challenges of the future with teachers by helping them interpret the changes going on and find meaning in their work, even as teachers at the same time help students learn and meet the higher academic standards. As the accountability norms, epitomized in the United States by the No Child Left Behind Act of 2001, threaten privatization of public schools as a sanction against failing schools, teachers, led by their organizations, have already recognized the dangers to themselves and public education if these trends continue. Thus, the harsh realities of work place conditions and sanctions for failure have already

---

Vignette 8.1

In a speech at J. E. B. Stuart High School in Falls Church, Virginia, January 26, 2005, President George Bush outlines a proposal to extend educational testing and accountability requirements to public high schools across the nation. "President Discusses No Child Left Behind and High School Initiatives," http://www.whitehouse.gov/news (accessed July 7, 2007):

> To ensure that the intervention programs are working and graduates are prepared, we need to be certain that high school students are learning every year. So the second component of my high school initiative is to measure progress with tests in reading and math in the 9th, 10th and 11th grade. Listen, I've heard every excuse in the book not to test. My answer is, how do you know if a child is learning if you don't test. We've got money in the budget to help the states implement the tests. . . . I've heard people say you're teaching the test; if you teach a child to read, they'll pass the test. Testing is important. Testing at high school levels will help us to become more competitive as the years go by. Testing in high schools will make sure that our children are employable for the jobs of the 21st century. Testing will allow teachers to improve their classes. Testing will enable schools to track. Testing will make sure that diploma is not merely a sign of endurance, but the mark of a young person ready to succeed.

awakened teachers to the reality of the big picture of the global economy. They are in it, and, potentially, have much to lose, as do other workers threatened with downsizing and privatization of their jobs, as do their students who will face a transformed world of work in the future.

Thus, historical changes, as in the past, have opened new possibilities for teacher consciousness of their work and of their place, as teachers, in the U.S. and global workforce. Their social class blinders will have the possibility of being removed. The resulting opportunity for class solidarity could reenergize the teacher union movement and reposition teachers as leaders in education and the labor movement. Thus, social class consciousness "counts" where it matters most, i.e., in empowering teachers to work together to meet the challenges of the new global economy at the work place, where its consequences can be the harshest, and where teachers, students, parents, and communities all stand to gain through their solidarity with one another.

References

Dolby, N. and G. Dimitriadis, with Paul Willis. 2004. *Learning to labor in new times*. New York: RoutledgeFalmer.

Shaffer, R. 2002. Where are the organized public employees? The absence of public employee unionism from U.S. history textbooks, and why it matters. *Labor History* 43 (3): 315–34.

Perlstein, D. 2004, *Justice, justice: School politics and the eclipse of liberalism*. New York: Peter Lang.

Carnoy, M., R. Elmore, and L. Santee Siskin. 2003. *The new accountability, high schools and high-stakes testing*. New York: RoutledgeFalmer.

Furhman, S. H., and R. F. Elmore, eds. 2004. *Redesigning accountability systems for education*. New York: Teachers College Press.

Lipman, P. 2004. *High stakes education, inequality, globalization, and urban school reform*. New York: RoutledgeFalmer.

CHAPTER 9

Campaign against the Opening of City Academies in England

Ian Murch

City Academies—later renamed Academies to allow their geographical scope to be extended—were introduced by then Secretary of State for Education David Blunkett in 2000. They were an important milestone in Tony Blair's program to bring about what he describes as an "irreversible" shift in state-provided education in England, wresting oversight of state-funded schools from the system that has prevailed for more than sixty years. Academies are removed from the control of the local education authority (LEA), parents, staff, and community representatives; from the local authority's range of support and intervention functions; and from LEA control of pupil admission arrangements.

The Plan

According to the plan, an Academy—sometimes entirely new, but more often replacing an existing school on the same site—would remain state-funded but would be handed over to a sponsor. It would have a new building—typical cost in the region of £25 million to £40 million—more than twice the money spent on other secondary schools (schools serving students ages eleven to sixteen). In return for stumping up "up to £2 million"—potentially in installments or even in kind—the sponsor would get the land, the new building, and control over every aspect of the school—how it is run, including how its staff members are employed, what the pupils are taught, and how children are chosen for admission to the school. Day-to-day running costs would be met in their entirety by the government.

The target was to have 200 academies up and running within ten years, but potential sponsors were actually hard to find in the first instance. Some who did eventually come forward have received peerages and knighthoods, and have reportedly been interviewed in the "cash for honors" investigation that began with investigations into political donations to the Labor Party. Most of the first wave of academies have been sponsored by rich individuals or the companies they lead, often being named after them, and sometimes developing a curriculum that reflects their religious or other worldviews. Bulk sponsorship is possible. Some businessmen are sponsoring more than one, but the sponsor planning to run the most academies is a Church of England (C of E) Company, called the United Learning Trust. For a relatively small outlay for the cash-rich but participant-poor C of E, this route will make it one of the largest providers of secondary education in the country.

Opposition to the introduction of academies was widespread when they were mooted. Most people involved in local government, together with teacher and parent organizations, strongly opposed this discount sale of public assets, and recognized how it could undermine equality of access to education, remove local democratic control and participation from schools, and allow particular interests—and even cranks—to impose their values on schools.

The government dealt with this opposition in the way it has typically dealt with opposition to its privatization and marketization proposals. It made only whatever concessions it had to make to ensure any rebellion in parliament did not prevent the passage of the legislation (easy given the support of the Conservative opposition) and then toughed things out.

Implementation of the Reform

This was, however, not the end of the story. Where an Academy is a new development on a new site, resistance is pretty difficult. But as we have seen, an Academy usually has to replace an existing school. This means there is the potential for the LEA, the governors, the staff, the parents of children already at the school, and parents worried about their access to their local school in the future, to all involve themselves in attempting to thwart the plans. There are many examples of this happening, and some of them have succeeded, despite all the determination of the Government to push through its "irreversible" changes.

Many attempts to create academies did not reach the stage of a public announcement of the plan. Many a visit by a Government minister or a Department for Education and Science (DfES) official to seek to persuade a LEA or a school governing body to go down this road has met with a direct rebuff. For this reason, a significant number of towns and cities with the sort of educational and social profile in which the Government would expect one or more academies to be created still have none. There are, however, pressures that can be exercised on reluctant LEAs. The most commonly reported one is the threat that unless the LEA uses its good offices to persuade a school to become an Academy, it will not be granted money for the refurbishment of other schools.

Once a named school is targeted and a potential sponsor identified, the work of resistance falls to the other interested parties. In Bradford, the LEA, in return for being admitted to a major school building and refurbishment program, agreed to give support to the identification of several schools to be turned into academies. This was apparently done by unelected officers, without elected members understanding what had been done on their behalf. When they found out, they expressed their discontent about being left out of the loop, but reserved their judgment about their attitude to specific Academy proposals.

Opposition Organized by the Union

Bradford is one of Britain's largest and poorest cities. It has one of the largest proportions of school pupils in the country for whom the medium of education is not their home language. Because of these characteristics, it has consistently been near the bottom of the league tables (published comparisons of school test results) that the Government uses to judge schools and education authorities. The New Labor neoliberal model leads to the downplaying of the need to address this underachievement by targeted extra funding or by coherent strategies and has instead led to a focusing on piecemeal privatization as a solution. The LEA had already been forced to outsource many of the services it previously provided to schools before its pressure for academies began.

Bradford has a big, well-organized branch of the National Union of Teachers (NUT) with a strong record of campaigning on issues of school funding, equity of provision and antiracism, as well as on the bread-and-butter issues for teachers.

The branch got the names of schools likely to be targeted, sent several briefings to its members about the implications, with news of the unfolding story, and met staff members—jointly with other unions wherever possible—to discuss what could be done. Governors and head teachers, when contacted about the proposals when they first became known, said that they had not been contacted prior to being identified. Most of them, as soon as they knew, made it very clear that they did not want to be involved. This made it very difficult to introduce potential sponsors, who would know from the outset that no one on the ground wanted them.

Two schools that were reluctant to be targeted were, however, particularly vulnerable to pressure. They were both in the English school category known as "special measures," due largely to below-average pupil examination and test results, which in turn are the result of the schools serving communities with substantial social deprivation and having the often consequent difficulty of recruiting and retaining senior staff. They are the kind of schools that the Government would regard as "ripe for takeover."

Serco, the private company that delivers school improvement services on behalf of Bradford Council, is not surprisingly a particularly keen advocate of new government initiatives that are unpopular with the vast majority of LEAs that continue to directly deliver their own support services. Serco has no democratic mandate or responsibility, just a contract to do what the government tells it. One of these government initiatives allows the LEA to seek the removal the governing body of a school in special measures and replace it with nominees appointed by the government. When the fine print of the contract between Bradford and its service deliverer was examined, however, it turned out that this power had been transferred to Serco. Soon after these schools had been identified as potential academies, the government was happy to work with Serco in sacking the governors.

The new bodies responsible for running the schools are known as Interim Executive Boards. Neither of them contained any parent or staff representatives. In one case, Rhodesway School, no one from the community or the previous governing body was included. In Rhodesway's case, the chair of the new board was initially Sir Peter Clarke, actually employed as the government's top man for implementing new initiatives. In the case of the other school, Wyke Manor, the chair of the board was the head of an existing Academy.

It took no great imagination to see what was going on here. There was a great deal of discontent among staff, parents and governors in both schools when they heard the news.

With the help of the NUT, a public meeting on the proposals for Academy status was arranged. Staff and parents from both schools attended, with other members of the public. The platform included an expert on education statistics who looked at the track record of academies, a speaker on how academies function and their power over admissions, and parents from a school in Doncaster, which had been taken over and turned into an Academy by the Emmanuel Foundation, the fundamentalist Christian organization set up by Sir Peter Vardy.

Parent and Teacher Response

To a dispassionate observer, it was shocking that the statistics presented by a respected academic researcher showed no more value added per pupil in Academies than at other schools, even though an awful lot of money had been spent that could have really been used to address underachievement in ways that are known to work. But to people asking "What's going to happen to me?" it was different things that created the shock waves that ran around the room.

Teachers were shocked to hear of the examples of schools that, despite the best efforts of unions, had imposed longer working hours.

Parents were shocked to discover that religious beliefs that they did not subscribe to could be introduced into a school their children already attended—introduced not just to religious education lessons and assemblies but also to science lessons and sex education. They were shocked to discover that their neighborhood school might create admissions criteria that denied their children admission and that there would consequently be no school for many miles in which they would be guaranteed a place.

The parents of children at the Trinity Academy in Doncaster then described their experiences. Many of us had heard the first Vardy Academy described as totalitarian by ex-pupils and staff and knew its record of expelling sixteen times as many pupils as its neighboring schools in Middlesborough. What we hadn't had was the raw emotion of parents put in the front line of this bizarre new kind of social experiment. "Every child has been given a Bible, and they get into trouble if they don't carry it around with them," one parent said. Another reported that her child had been excluded for allegedly having gray trousers bought at the wrong shop. They were in fact trousers bought at the right shop, she said, but out of a new batch. She described the impossible task she faced when trying to explain this and get an apology for her child. A third parent said it was pointless complaining because the head just said, "If you don't like it you can go somewhere else." She described a get-tough policy that denied children the right to go to the toilet during lessons, however pressing the need. There had been 100 disciplinary exclusions in the first term, we were told.

In summing up, one parent said, "We feel that the school is trying to get rid of kids who have problems or won't do well in exams or even question things, They've got a waiting list with kids from outside the area who want to go to this brand new school, and they want to get them in so the exam results go up. But what happens to our children—the ones they don't want?"

Following lively discussion, questioning, and the planning of future activities, the meeting was about to end when one of the audience identified himself as the chair of Bradford LEA's Education Scrutiny Committee. "My eyes have been opened," he said. "We've looked into Academies once, and we thought we knew they could raise standards, but that investigation was dominated by people from the DfES (Department for Education and Skills). I'm going to reopen the scrutiny of Academy proposals, and invite all the speakers at this meeting to give evidence to the councilors on the panel."

He was as good as his word. Councilors of all parties and co-opted members of the Scrutiny Committee spent a day in a wood-paneled and stained glass windowed room of Bradford's magnificent Victorian City Hall listening to this evidence. They questioned the academics and trade unionists closely and vigorously.

They did not in the end come up with outright opposition to academies. Their report strongly criticized the government for claims about the Academies program that were not supported by any evidence and made many other telling points, including the following: "It would seem that the DfES has the ultimate say over which sponsor is to be used for a particular new Academy. Given that any new Academy is a local solution to a local problem and that certain sponsors want to introduce particular approaches and beliefs to the running of the Academy, it seems bizarre to the committee that the choice of sponsor is essentially made by civil servants in London."

Those of us present felt that their financial dependence on central government contributed to councilors' reluctance to make a formal declaration of opposition to a pivotal government policy. However, they did set numerous criteria they expected for any Academy proposals that contradicted many of the aims that the Government had for the program, including the following:

1. That the agreement between the DfES and Academy Trusts enables the DfES to remove a sponsor in a direct way if the sponsor proves unsuitable or fails to meet their liabilities under the agreement and that the Local Authority can make representations to the DfES if it believes that the sponsor is proving to be unsuitable.

2. That, in order to establish a sponsor that has the support of the community, the choice of sponsor is made in full consultation in an open and transparent manner with the relevant stakeholders of the failing school.

3. That in order to ensure that Academies are really accountable to parents, the number of parent governors and teacher representatives on the governing body of an Academy is determined in the same way as the number of such governors on the governing body of a local authority community school.

While this report was being written, the other interested parties were active. A government minister came twice, without the knowledge of local councilors, to try to introduce sponsors and pave the way for the transition to Academies in the two targeted schools.

No one wanted one of the schools. It lies in the shadow of one of Europe's major producers of hazardous chemicals. As a result, it is now nowhere in any part of the government's program for refurbishing secondary schools and looks set to remain in its run-down and demoralizing buildings unless its staff win their fight for something better.

A consortium of rich businessmen with no connections with Bradford was found to make a bid for the other school. Parents and teachers who had attended the original public meeting quickly formed a campaign group. They fought for and obtained representation on the interim executive board that runs the existing school. They produced material setting out their concerns about the proposals and held a series of meetings.

A teacher from the school takes up the story:

> After the initial shock, all staff were invited to a meeting hosted by all the teaching unions and Unison (the largest non-teaching staff union). It was agreed to hold a short demonstration, on an open day, outside the school gates to which the local media were invited. There was a great deal of pressure for staff not to take part in this demo, but following an extremely well attended, voluntary staff meeting it was decided to go ahead with the protest and that a committee should be established to co-ordinate opposition to the Academy, ensuring all concerned would be consulted—staff, parents, and students all of whom the sponsors had forgotten up to this point.
>
> As a consequence we eventually got to meet Lord Bhattia and friends. They told us how they were going to turn everything round and rescue the school, but produced not one piece of concrete evidence of how they were going to achieve this apart from spending public money handed to them by the present Government. The Labor Government promised to provide every school with new buildings. We shouldn't have to put up with our school being taken away from the local community to get this.

A Story with No Ending

This is an unfinished story, because at the time of writing the campaign continues.

Rhodesway School has come out of special measures; has strong leadership and a stable, well-motivated staff; and has been judged to be one of the 200 most improved schools in the country by the government's own education ministry. Its examination results have improved well beyond targets, and well beyond results obtained by most existing Academies. In some ways, the desire to fight off the threatened Academy has helped because it has brought the school community together. The resolution of the staff, and of the Parent Teacher Association, is, if anything, even greater now to protect "their school" from an outside takeover. Yet its improved results make it paradoxically a more attractive target to the would-be new owners.

The Rhodesway story is a picture repeated across the country, as a recent House of Commons inquiry revealed. The teacher and parent governors from Rhodesway joined representatives from more than thirty local campaigns to prevent the imposition of academies who attended to give evidence. A vibrant national organization—the Anti Academies Alliance—has now been set up to coordinate such resistance and to make the educational and political case for genuinely comprehensive education under local democratic control.

We cannot say yet what the actual outcome will be of the Rhodesway campaign. What we can say is that the campaigners by their actions will win something. They will either keep their school as a genuine community school, or they will secure the concessions they have already won over fairer admissions, staff and parent involvement on the governing body, and protection of staff conditions and rights.

The community that constitutes a school is a very powerful force for mutual good when it mobilizes itself. When it comes together with other such communities, it becomes the engine for a movement that can give real force to the reassertion of egalitarian and democratic values against the neoliberal agenda.

Vignette 9.1

From James Tooley, "Welcome to easyLearn, Class 1," *The Times* (London), April 17, 2006:

> . . . Gazing into my crystal ball, I see chains of learning centres carrying the distinctive bright orange logo of "easyLearn," competing with those sporting the red "V" of "VirginOpportunity."

CHAPTER 10

An Inner-City Public School Teacher's Story from China

Yihuai Cai

Two years ago, when traveling back to my hometown of Hangzhou, a cosmopolitan city in southeast coastal China, I was stunned by the sight of a brand-new glittering title hanging outside my former elementary school—Qiushi Education Corporation. My old school, a former key district school[1] receiving adequate government support in terms of funding, facilities, and teachers because of its better performance and reputation, has been replaced by a corporate venture. Qiushi Elementary School has expanded its franchise to include three newly established public schools in other districts (*minban* schools).

Education corporations are no longer a rare sight in China's developed urban areas. Over the past four years, Qiushi Education Corporation is only one among twenty-eight education corporations that have emerged in the coastal city of Hangzhou. According to the city government official website report, these corporate schools often include one key public school, other affiliated newly established public or private schools, and corporate or joint venture–affiliated schools. The ownership of the public schools is still in the hands of the government, but their administration is contracted out by the education department to enterprises, business organizations, social organizations, or individual investors. The local education ministry suggests that the consolidation of schools and the transformation of the management system of the public elementary and secondary schools is a necessary strategy toward education equality and the fair sharing of educational resources in an efficient manner. Corporate schools claim to be able to mobilize more funds and human resources, create their own management styles, improve teaching and curriculum on a larger scale, and arguably offer city residents more affordable public quality education.

However, Z, a young woman who has worked in the same public inner-city middle school since graduating from her local teachers' college in 2001, told me a very different story from the one described in frequent headline references to the new schools' "quality" and "choice." Z and I have been in regular contact regarding subsequent educational developments within the city of Hangzhou. The middle school in which Z teaches does not fit so neatly into the blueprint for the city's education landscape. Because the political situation in China makes criticism of the government dangerous, in the following description, I use Z's initial when I describe what she has witnessed and said. Her quotations are translated from our online and in-person conversations in Chinese.

Aside from the administrative order to consolidate with a nearby middle school in 2002, nothing much has changed for this inner-city public school. Located amid an urban landscape that juxtaposes collapsing old residential apartments and rising modern business skyscrapers, Z's middle school epitomizes the awkward situation many such inner-city public schools face in the midst of ongoing market economic change. At a time when new school facilities are being built in the upscale suburbs to which most residents have moved, inner-city schools are left to prove their own quality and develop their own survival strategies.

"It is said that the local district will allocate funds in an equal manner, but in recent years, our school seems to be getting less and less resources," Z told me, expressing her worries. "We are in no position to compete with those rich private schools already, and the local education ministry seems to be busy with managing those education corporations. We are very much neglected in the corner. Maybe the plan is that one day when the central business district is fully established, schools like ours will be closed down."

Despite its relatively small scale in comparison to the emerging giant education franchises, Z's school is serving an important role for the population trapped in the inner-city. Many of the children attending the school have families whose incomes have suffered because of the ailing state-owned enterprises (or other such economic vulnerabilities). They cannot afford new apartments in the suburbs. Some families are rural migrant workers who can not afford the high entry fee to attend other bigger schools. In Hangzhou, as in many other Chinese cities, attending schools outside one's residential area requires paying a significant entry fee, a one-time personal "endorsement" to the schools besides the yearly tuition.[2] Z's school charges a U.S.$100 entry fee per student, a comparatively cheap amount in the city. Although officials claim every family should enjoy the same educational opportunities, families that cannot afford to live in more expensive areas or pay higher entry fees have limited educational opportunities.

"It is never a fair game for schools like ours. Teachers from my school are often overworked to educate those children coming from poor and illiterate families. But the measurement of our work is always the same—how our students have scored compared to those wealthier franchise public schools in the yearly city-wide public school quality examination and the provincial standard high-school entrance exam." Z recalls her frustration with the annual city-wide quality examination: "Every year the city education ministry will select twenty-five public schools to participate in this quality examination. Despite the central government's advocacy for all-round development, the content of this examination has always been based on two subjects only: literacy and math." What's odd is that only scores from students in the bottom third count as the measurement for schools' education quality. The seven schools with the lowest overall score receive a public warning. Z says this is truly a torture every year for both the teachers and the students. She explains,

> This year, the 9th grade in our school was selected. Teachers teaching that grade had to teach on average six to seven classes every day for a whole week before the test. Only two subjects are taught during that whole week. And of course, they are literacy and math. Every year I will have my fingers crossed that my class will not be selected as the quality sample. My teaching bonus and more importantly the ultimate judgment of my teaching are both tied into this unfair and time-consuming activity. The exams are designed by the city or district education researchers who rarely travel down to our classrooms. What I can't understand is how so many education experts claim this to be the most efficient measurement of a school's quality when they don't know what teachers are facing on a daily basis. As a teacher, I feel my hands are tied and I have to worry constantly about how my students perform in these tests.

Besides the simplistic measurement of both teachers' and schools' quality, a more profound issue that contradicts the official claim of fairness and equality within the trend of consolidation and marketization of schools is the well-being of all teachers. In 2002, when Z's school received the city government's order to consolidate with a nearby middle school, the school administrator was told to maximize the efficiency of managing school resources by reducing the number of teachers, meaning many teachers from these two schools had to be relocated. Z commented, unsettled at the possibility of relocation or severed compensation, "It is as if overnight you will possibly become a substitute teacher after all these years of teaching."

There is no official layoff policy. Z explains that "usually teachers will face two options. Either they agree to teach in some newly established public school on the outskirts of the city—you know, those really far away districts where other teachers don't want to go. Or they can choose to stay in the school teaching relatively unimportant courses and with significantly reduced teaching assignments. But then, of course, the salary will also drop significantly, because our salary is directly related to the teaching loads."

I was startled and saddened upon learning the average teacher's salary at Z's school. As a diligent teacher working over ten hours every day at school with full teaching load, Z receives around U.S.$120 (900 yuan) for her base salary every month plus some U.S.$90 (600 yuan) teaching bonus based on her teaching load. As little as this U.S.$210 monthly salary may sound, Z told me that hers was among the highest in comparison to other teachers. "For some teachers, they can only receive a total of $150 (1100 yuan) every month. That's why we often joke that we are among the poor who need to be on subsidies!" In a city with a booming economy and ever-rising real estate prices matching the U.S. standard (around $70,000 for a small, two bedroom flat in average residential areas), the treatment of its teachers is heartbreaking.

"I can't imagine what will happen if I get pregnant. I've seen too many women teachers losing their teaching bonus once they are on their maternity leave and having to live on a meager monthly salary."

Stunned by the current working condition of Z and possibly many other inner-city public school teachers, I inquired about the existence of any teacher union that protects the teacher's interest. After hearing my description of the teacher union in the United States, Z explained to me the different situation in China. Teacher unions in China are not independent. Rather, they are organized at different government levels. Take Z's school, for example. The chair of the teacher union in her school is a full-time, school administrative status position (normally the equivalent of school vice president). The chair is usually selected among several administrative candidates by teacher representatives in her school and then appointed by the school president. The school chair then will represent the teachers in his/her school at the district teacher union meeting.

Z is a teacher representative this year and has just attended the school's yearly teachers' representative meeting:

> Usually we have a preliminary meeting before the formal one to inform us about the major issues that we need to discuss during the meeting. For example, this year the chair told us it was about the establishment of the tenure system.

Although teachers have different opinions, this proposal still got passed almost unanimously, because we all sort of know this proposed issue most probably has already been decided among the administrative or even higher level officials.

Of course, sometimes teachers also raise proposals, such as asking for higher salaries or expressing dissatisfaction with the school's decision to lease out its property to companies. . . . But for most of our proposals, we seldom hear any feedback or updates on the progress. To be honest, besides the one hundred yuan [around fifteen dollars] each representative can receive for attending the meeting, I don't think there is much the school administrators can really do, because to solve many of these issues is beyond the school's capacity. . . . I don't know whether the issues we raised during our school's meeting ever reached the district or city level.

An independent teacher union that protects and speaks for teachers' interests seems especially important for those inner-city public school teachers, whose voices rarely reach higher official levels. How such a teacher union can be carried out in the Chinese context is another important and complicated issue that needs more discussion.

Toward the end of our interview, Z sadly recalled her twelve- and thirteen-year-old students' comment that the best they can expect in life is to be a cab driver. "As a teacher, I want to make a change and plant hope in these poor young children's hearts. But for now, I can only do what is asked of me from my own integrity as a teacher, and I don't know how long I can hold onto that with my meager salary and limited opportunities to be the teacher I have always wanted to be."

Notes

1. In the city of Hangzhou, there are 12 major key elementary schools and 6 key middle and high schools located in the six districts. The system of key district schools was established more than twenty years ago in China as an attempt to allocate limited resources to those schools with better performance. The purpose of the policy during that time period was to establish some sample schools for quality education. In recent years, the naming of "key schools" was abandoned by the federal and local government in favor of a discourse for education equality. However, I have observed that most of the newly established education corporations are former key schools that have been receiving additional government support over a decade.

2. In China, *hukou* (the household registration system) is still in place in terms of managing the massive population. Depending on one's birthplace, a person can be a rural or urban resident, receiving different education, employment, and other benefits and treatments. In the city of Hangzhou, as in many other urban cities, students usually go to nearby elementary and middle schools based on their residential area as recorded on *hukou*. Students who want to enter a public school but live outside the school district have usually two possible courses of action. The family can either purchase an apartment or house within the

school district to qualify as district residents or pay an entry fee to that school. The amount of the entry fee is decided by the individual school based on its quality and reputation. Usually former key district schools charge a much higher entry fee than other public schools. In Z's case, because her middle school is not among those key schools, her school charges a much lower entry fee for students who live outside the school district.

CHAPTER 11

What Teachers Want
from their Unions
What We Know from Research

Nina Bascia

N ever before has organized advocacy been more important for teachers (and for public schooling more generally). Yet there are few
teachers' organizations anywhere that can honestly claim broad, sustained, and positive relationships with their members. Teachers' perceptions
of their unions are obscured by a lack of organizational visibility, by a focus
on an apparently narrow range of issues, by a persistently negative and unflat-
tering press, and by a sense that the union "belongs" to only a select group of
teachers. These factors have been in play for several decades, but the distance
between unions and teachers has increased in recent years, for several reasons.

First, at the current time, as a large generation of teachers has retired, with
them has gone the personal, visceral appreciation of the importance of union
advocacy, and unions around the world are grappling with how to produc-
tively engage with a new teacher cohort that lacks this understanding.
Second, the centralization of educational decision making—restrictions on
collective bargaining and teacher involvement in decision-making arenas
beyond their own classrooms—has resulted in fewer opportunities for teachers to become involved in, or at least aware of, union advocacy with respect to
quality working conditions. Third, negative comments by governments and
media about teacher unions, and many unions' counterattacks, have increased
in frequency and intensity, making many teachers uneasy or ambivalent about
identifying with them. Fourth, as public education and teachers have come
under governmental attack, many unions have been overwhelmed by the

magnitude of issues in need of addressing, and there is a common tendency for them to narrow the range of concerns they claim to stand for. All of these factors seriously confound teacher unions' viability: their legitimacy as the voice of teachers, the breadth and quality of information they base their actions on, and their capacity to meet the range of teachers' needs and wants. Given global trends that put public education at risk, teacher union viability is more of a cause for concern than ever before.

What teachers want from their unions is both an individual and a collective phenomenon. To some extent, what teachers want depends on who they are, where they are, and the kinds of students and programs with which they work. Most descriptions of teacher unionism view teachers as a homogeneous mass or, at best, as several distinct factions. They are seen as either only capable of "selfish" self-interest or are easily manipulated by personally ambitious union leaders. Closer glimpses of what teachers value reveal tremendous variation in what they want from their unions and why, and yet there are patterns to this variation. The structural and political conditions of teaching engender certain kinds of needs and wants: the ways social class, race, and gender play out in schools; teachers' subordinate status relative to administrators and policy makers; and the challenges inherent to teachers' professional identity encourage teachers to turn to unions for resolution and yet, paradoxically, often to have an ambivalent relationships with their organizations. When teacher unions recognize these patterns and take them into account, they can greatly increase the power of unionism for teachers and the power of unions in their advocacy roles on behalf of teachers.

This chapter draws from nearly two decades of my research—from the late 1980s to the middle of the first decade of the twenty-first century—to describe the patterns inherent to teachers' opinions of and involvement with unions in Canada and the United States. These studies took place in Alberta and Ontario, California, New York, North Carolina, Ohio, and Washington State. Some documented the value of unionism for a wide range of teachers (Bascia 1994); others concentrated on the work of union activists (Bascia 1997, 1998a; Bascia and Chassels 2004; Bascia and Rottmann 2005); many others concentrated on teacher union efforts to promote positive educational change (Bascia 1998b, 2004, 2005; Bascia, Stiegelbauer, Watson, Jacka, and Fullan 1997). My assertions are based on patterns of responses across many teachers and in many different sites. The quotes are drawn from interviews with teachers in these studies. While the evidence from which these assertions are drawn is firmly North American, there is likely to be some resonance for educators elsewhere in the world.

How Teachers Differ

There are several common claims about what teachers want from their unions; all of them are accurate some of the time. Educational policy research, which has tended to steer entirely clear of teachers' perspectives on labor, maintains that the "rank and file" is only interested in material issues and responds with suspicion when union leaders express an interest in reform (McDonnell and Pascal 1988; Olsen 1965). There is very little research on how teachers view their unions; but the evidence that exists suggests that they run the gamut from strong antiunion sentiments through apathy to persistent and passionate union affiliation. Teachers' opinions and actions about the value of what their union has to offer are rooted in their own "common sense," which we can see as a product of complex sets of personal and local factors that, taken together, produce particular occupational, social, political, intellectual, and economic needs and values.

Teachers can find themselves faced with situations where they believe they have not been treated fairly and are unable to teach as effectively as they want to. This can happen to an individual (one teacher treated differently from another in the same school), to teachers collectively (for example, where many teachers in a school or school district believe they have been "treated like children"; Bascia 1994), or where administrative or policy decisions have been taken that compromise their ability to teach well. On the other hand, some teachers are well looked after (or can look after themselves) and will not necessarily recognize the value of advocacy ("Why would you need association protection for your job unless you have hassles? And there are so few things that you can be hassled about in teaching"; Bascia 1994).

The extent to which teachers' understanding of the value of union advocacy is an individual or a larger, shared phenomenon also depends on whether, in a given school or district, teachers' work is largely isolated and independent or, on the other hand, teachers have opportunities to develop personal and professional relationships that allow them to see the value of union advocacy for their colleagues—for example, a male teacher in a socially tightly knit school recognizes that "teachers get desperate. . . . To be raising a family, to have a wife who may not work, to want to educate those children . . . and realize that another district will pay more, you become angry" (Bascia 1994). There are important variations in how teachers' sense of professional community (and where teacher unionism fits) play out in different school contexts (Bascia 1994; Little 1992; McLaughlin and Talbert 2001). In some schools, union affiliation is generally seen as a positive component of staff identity; in other schools, it's up to the individual teacher to choose whether or not union identity "suits me."

Teachers work in different environments—urban, suburban, or rural; in inner city or more privileged neighborhoods; with transient or stable student populations. These community settings create different needs and different degrees of available capacity for teaching and learning (Bascia and Jacka 2001). They play out across districts and schools and within schools, across different programs and tracks or streams. Teachers who work with non-native English–speaking immigrant children, children with special needs, or any other population of students whose academic success is not easily achieved have greater need for access to support, information, and influence. As Chris Stewart observes in her essay, they are also least likely to get support without some considerable effort on their part, because teachers' status within a school is linked to the status of the students with whom they work, as well as the subjects they teach (Bascia 1997; Finley 1984; Little 1993; Siskin 1994). Teachers in higher status roles (for example, teaching in science) may also be union active: a sense of entitlement, an expectation that resources should and must be made available for their programs, and a sense of outrage when they discover how teachers are excluded from domains where decisions about resources are made drive such teachers to union activism. Union leadership can be a poor (wo)man's version of administrative authority—for example, a male union representative described his working relationship with a male principal as a contest of equals: "We would both get magenta in the face screaming at each other and then go have a glass of wine" (Bascia 1998a). These are some of the most obvious examples of how gender and other social status markers play out in terms of union involvement and acceptance. Teachers between the two status extremes, all things else being equal, are less likely to be involved.

Teachers' own social class, gender and race create different realities for them. These realities may engender particular goals and commitments and, further, these social identities may result in expectations by educators, students, parents and others that they perform certain roles and not others. That women and racial minority teachers have faced greater difficulty in achieving administrative leadership positions than white men and the expectation that minority teachers assume special responsibility for special needs students are two examples of the ways that social identity shapes teachers' work and, in turn, the needs and desires teachers have for organizational advocacy or union involvement (Bascia 1997, 1998a; Bascia and Jacka 2001).

Another factor that complicates the answer to the question "What do teachers want from their unions?" is that teachers change over the course of their careers. Their skills as teachers grow, and their need for on-the-job support shifts. What piques their interest in terms of learning and taking on new challenges changes over time. As their bodies age, their energy levels change. Their personal lives may alter over time as responsibilities for family ebb and flow. Why and when teachers become interested in union involvement changes as a result: younger teachers are more likely to be attracted by high quality teaching-related professional development, more seasoned teachers by the availability of interesting new professional opportunities and the chance to develop leadership skills. (These avenues are particularly important for teachers who are not interested in, or have not been successful in, achieving administrative positions—teachers viewed as, or who view themselves as, unconventional, and teachers interested in promoting educational ideas that may not be at the top of policy makers' lists.) Teachers of different generational cohorts will have experienced different occupational histories: eras of particular reform ideas, professional opportunities and constraints, and union prominence or invisibility.

All these differences make it very challenging for teacher unions to select organizational priorities and goals that will be meaningful to all or even many teachers. Failing to recognize the roots and complexity of teacher diversity, many union leaders search for organizational initiatives that offend no one but end up satisfying few.

What Teachers Want

Despite the complexities described above, we can glean from research an understanding of what teachers want: occupational advocacy, economic sufficiency, participation in decision making, professional development and learning opportunities, and the articulation and promotion of a positive professional identity. At first glance, these desires appear to match unions' typical priorities—but as this and the next sections suggest, it takes a deep understanding of what teachers want in order to do them well.

Occupational Advocacy

Because teachers generally have few opportunities to make decisions about the conditions of their work, they must look to unions to speak with administrators and policy makers on their behalf. Teaching assignments, class size, daily work requirements, health and safety provisions, evaluation strategies and their consequences, teaching resources, available professional learning opportunities, and the requirements for professional advancement are all critical to the nature and quality of teaching—and all subject to administrative and governmental discretion. The hierarchical and bureaucratic nature of educational organizations and systems results in gaps between what teachers say they need and what policy makers and administrators believe is best. People in positions of authority over teachers often overlook the particular details that make a difference for teachers (Bascia 1994, 1997, 1998a). Teachers have little appreciation for actions taken that don't correspond to their occupational needs. A leader's enthusiasm for a particular reform is met with suspicion by teachers who don't see its utility. Effective teacher advocacy is often a scarce commodity because of the variety of teachers' lived experiences in teaching.

Economic Sufficiency

Working conditions, salary, and benefits are fundamental to which and how many individuals are attracted to and remain in particular occupations and settings, so it is ironic that politicians and the press often view teacher unions' concerns with compensation as "self-interested" and "nonprofessional." Salary and benefits are persistent points of contention in labor relations because of the low priority placed on educational spending by many governments, and they represent the "necessary conditions" for teacher quality (Bascia 2004). For teachers, they have both real and symbolic value: teachers' ability to focus exclusively on teaching rather than having to take on additional work for wages to adequately support themselves and their dependents has a real effect on how well they teach. Beyond this, the fact that, legally, unions can only negotiate about salary, benefits, and working conditions means that other issues and concerns get funneled through these particular and narrow channels. Demanding a salary increase often symbolically stands in for teachers' frustration with policy directions and their concerns about not being treated respectfully (Bascia 1994, 2004; Carlson 1992).

Participation in Decision Making

Teachers want their perspectives taken into account when educational decisions are made because, as suggested above, it can make a profound difference in their ability to teach well. Many are content with a representative system of participatory decision making, given the extensive time and energy demands of teaching, but some desire personal involvement. Teacher unions provide several venues within their own organizations and through representative decision making structures involving administrators and others in schools and school districts. A teacher's desire to participate is often driven by her sense of responsibility for a particular student group or program: "[My program is the] bastard child of any curriculum. Math and science have a real good valid reason for funding, but we have to really scream and holler to get funds"; "I try to be a part of the group as much as I can be. . . . I feel that if I'm respected then [my special needs] students are going to be more respected in their classes in the mainstream" (Bascia 1994, 1997). Participation in decision making may also be tied to an understanding of teaching as a commitment to children that by definition must extend beyond just classroom teaching: "[As] much as me being here and working with them during the time that they're here, [it's also] a commitment that I try to fight for smaller class sizes, that I try to fight for more aides, that I try to fight for better teachers. . . . It's naïve to think 'I'm only here for the kids, I just want to focus on my classroom and the kids and ignore the rest of it'"(Bascia 1997).

Professional Development and Learning

Many teachers want to grow intellectually and to develop their teaching-related skills over time, and sometimes their unions are logical venues for professional learning through structured workshops and courses on topics as diverse as teaching strategies, student assessment, and school-based decision making, and they also provide direct assistance to "floundering teachers"—new teacher induction, peer review, and special supports for more seasoned teachers in danger of being found professionally incompetent (Bascia 2000). Union-provided professional development has increased in frequency, variety, and visibility in recent years, partly because unions want to demonstrate that they take responsibility for teacher competence, partly because they serve as incubators for educators' innovative ideas, and partly to compensate for the increasing narrowness of school systems' professional development programs (Bascia 2005). Teachers want high quality learning opportunities, and many like the idea that their own professional organizations are the providers, because it is consistent with notions of professionalism.

Particularly at times when they are developmentally ready to try something new and expand their horizons as teachers (Bascia and Young 2001), participating in decision making or developing or coordinating new projects can give them opportunities to learn new skills, develop more complex understandings of educational practices, and participate in larger and different educational arenas. The exposure and experience these activities provide enable teachers to move beyond the taken-for-granted familiarity of their own classrooms and schools and come to understand the familiar in new ways. They allow teachers to engage with new ideas and with other teachers in ways that may not be possible within the logistical and social constraints of their own workplaces (Bascia 2000; Cochran Smith and Lytle 1992). A union-active teacher describes her career: "When I first came into teaching I was a classroom teacher . . . then I became head of a department and learned how the school worked together as a whole. And now through the union meeting I learn how teachers from different sites work together. So I'm getting a wider view of the coordination between people at different levels" (Bascia 1994).

Articulating and Promoting a Positive Professional Identity

When perplexed by the obstacles in the way of their daily or career-long professional efforts, teachers appreciate when their union provides the analytic concepts that help them understand their experiences as part of a larger constellation of systemic relations. Even better than mere abstract concepts are empathy, support and concrete opportunities to join in organizational actions that challenge systemic inequities. For example, a woman elementary teacher in Ontario noticed that her own opportunities for administrative advancement differed from those of male teachers: "I was watching my male colleague being tapped on the shoulder and told, 'You need to get into leadership activities.'" When she applied for a leave to work on a graduate degree, she discovered that only male teachers were granted an additional percentage of their salary for each child: "Can you believe this, this was in the late [1970s], this wasn't 1902." She sought federation representation and was provided not only with legal support but also an orientation to gender issues that helped her understand and articulate her experiences: "By going to [federation] meetings I was learning about the discrepancies, that there was not equity in terms of treatment and that these were sexist behaviors and unacceptable. That was very critical to my awakening, I think" (Bascia 1998a). She became active in her local federation and eventually provincial federation politics.

Teachers want to feel pride in their collective identity. When disparaged by politicians, administrators, or the press, they not only want someone to defend them, they want someone to challenge the prevailing antieducation, antiteacher rhetoric by providing alternative, positive, persuasive images of teaching and schooling (Bascia 2008). When their issues are belittled or misunderstood, when they are portrayed as simply selfish and self-interested, teachers want an effective way to challenge and recast public understandings about teaching and schooling. Teacher unions can do this through their public relations work; public relations is a common and increasingly active organizational function in teachers' organizations.

Teachers do not want to be embarrassed by the messages conveyed by union leaders claiming to represent them, though what embarrasses one teacher may not embarrass her colleague. Gender often plays out here, in the ambivalence teachers feel about the feminine coding of their occupational identity: for some, authority is conveyed through a show of masculine strength—being able to stand up to the big men in formal positions of power (for example, a long-time female union activist said, "When [the current president], who's big and appears tough and has a deep voice was elected, it was so easy for him to develop that level of credibility. . . . He made my job a whole lot easier" (Bascia 1998a). Others are uncomfortable with bellicosity and overt "political" behavior. The teachers' position—subordinate to administrators and policy makers, where their very identity and credibility is easily challenged—makes the politics of legitimacy complex and difficult.

What Teacher Unions Can Do: Strategic Organizational Choices

Many claim that teacher unions' downfall is their inability or refusal to act like "professional" organizations. What teachers want from them is certainly bound up in their own status as "not professional"—but clearly, whatever teachers or their unions do may be subject to charges of unprofessionalism since the concept itself is murky, and used more often as a weapon than a helpful guide (Bascia 1998b). Unable to assert authority over the conditions of their own work or to participate officially in the development of their shared craft, they may turn to their unions to try to make up for or at least minimize the difference between their occupational needs and goals and the realities they face daily. Unfortunately, many teachers do not perceive their unions as credible, positive and effective advocacy organizations either for themselves, their students, or the educational system itself. Except in times of real crisis or in the rare instance where the teachers in a district, school, program, or department share the conviction that union involvement is "part of who we are" (Bascia 1994, 1997), teachers tend to appreciate their unions in individual and idiosyncratic ways, based on personal experience and desires.

Unions themselves must take some of the responsibility for the extent to which teachers view them as viable organizations. Teachers' ambivalence, apathy, and frustration—with decision makers and with unions themselves—are rooted in part in the strategic choices unions have made about their relationships with their members, their internal organization, their strategic directions, and the discourse about teachers and teaching they promote publicly.

Many unions have the reputation among teachers of favoring, or "belonging to," one type of teachers over another—elementary or secondary, science or special education, one generational cohort or another, men or women—because of the patterns they see in who is elected into leadership, who participates in union activities, and what kinds of issues become organizational priorities (Bascia 1998a, 2000). Such situations tend to perpetuate and intensify, not only because teachers' perceptions of favoritism encourage or discourage them from active engagement but also because teachers who do have access to organizational decision making lack the awareness and information that might influence them to increase opportunities to union access and union priorities. Such situations can lead to uneven commitment from teachers, leadership vacuums, antagonism, and much energy expended in internal power struggles.

Most teacher unions are organized internally into discrete units, much like the departmentalized structure of high schools. Staff members associated with collective bargaining, professional development and teacher welfare typically work independently of those in other units, are responsible for distinct tasks, tend to interact with distinctly different people outside the organization, and as a consequence maintain distinctly different views of the world. They may intentionally or unintentionally hoard information (sometimes not knowing how what they know might be useful to others outside their unit); they often compete for resources and organizational influence; they may even work at cross-purposes. Collective bargaining and professional development staff members often find themselves in this kind of diametrical opposition. Their limited and particular views of the world result in less effective, less resilient programs and strategies than if they had shared information and expertise. Their independence from one another leads to organizational fragmentation and incoherence. Educators viewing such efforts are often frustrated by what they perceive as inadequate responses to their requests for support.

Confronted with a diversity of goals and values among members and unskilled at managing conflict, union leadership may become autocratic and authoritarian. Participatory decision-making structures become mere window dressing, their agendas and procedures manipulated, while real decision making rests in the hands of a small number of people or even an individual. Squashing or hiding conflict, of course, does not get rid of it, but it does send a signal to teachers that the union is not accessible to everyone, and it reduces the information and ideas available to staff members as they do their work on teachers' behalf.

Facing hostility from the educational system and a growing number of demands from their members for support, many unions adopt a "triage" approach, choosing to mount a small number or even a single agenda priority in order to ration scarce organizational resources (Bascia 2005). But focusing on a narrowed agenda, like securing salary increases for teachers, resisting reform, or demonstrating "reform mindedness" by promoting a single educational innovation, usually backfires: both teachers and others instinctively perceive the inadequacy of the vision driving the agenda, and many teachers are left, once again, with the perspective that their organization is not interested in promoting their best interests.

Many teacher unions, attacked by politicians or administrators or attempting to counter attacks on their members, respond in kind, adopting the language and terminology being used against them. For example, in Ontario in the mid-1990s, after several decades of cooperative relations between the province and teachers' federations, when a new government reduced the scope of federations' authority over teaching-related issues and began calling them "unions," the teacher organizations decided to "play hardball" and adopted the "union" moniker. In doing so they also adopted a narrow definition of teacher unionism and its occupational and social responsibilities. Teachers' organizations participate in defining the public discourse about teachers and teaching by helping shape the terms of teachers' work through collective bargaining, through communication with teachers and administrators, and through statements they make in the press. They can reinforce or challenge images of teachers as victims or heroes, passive dupes or active agents, technical or intellectual workers, or political activists or professionals (Bascia 2000). When unions respond to attacks that cast teachers as selfish and teaching as technical work with arguments that fail to contest these characterizations, teachers are handicapped in challenging negative press.

Many unions, then, inadvertently reinforce the status quo: they participate in keeping teachers alienated from the decisions that affect their work. Even while some teachers are actively engaged, many are not and cannot see the union as a vehicle for positive educational change. Even while a union may be engaged in multiple activities, its actions may run at cross-purposes to one another. While innovation may be occurring in various parts of the organization, its effects are not far reaching. And even when the official messages emanating from union leadership are "revolutionary," they may not reflect or be taken up by many teachers.

What Lies Ahead?

As contributions to this book indicate, change is afoot internationally. Within Canada and the United States there are a small number of teacher organizations that demonstrate that the problems in unions' responses to attacks on teachers and teaching need not always be the case. Their sustained viability in the eyes of many of their diverse teacher members is not just chance and not just a matter of charismatic leadership; they are strategic in their thinking about membership participation, organizational structure, leadership, and priorities, as well as how they articulate the value of teachers and teaching.

Viable teacher unions make a point of providing a range of different ways that teachers can participate in their organizations. Rather than emphasizing an orthodoxy in terms of the kinds of activities they sponsor, they make member interest and access a priority—for example, providing a wide assortment of different professional development formats and topics, scheduling and locating them in ways that make them accessible to busy working teachers. They provide a range of leadership opportunities so that many different teachers can develop organizational skills and become involved and known to others. They make a point of rotating the demographics of leadership so that neither teachers nor outsiders develop the impression that the union is not representative of the broad teacher population (Bascia 2006).

Viable teacher unions attempt to minimize balkanization and fragmentation by ensuring that staff and elected officials have overlapping portfolios that put them in contact with many different people and issues both inside and outside of the organization. They work to ensure that different units are not competing for resources or prestige. They develop initiatives that require people to work across units and demand many different kinds of expertise to bring to fruition.

Staff in such unions share information and borrow ideas from one another regularly. They spend lots of time in the field, interacting and sharing information with others not only in different sites but up and down the educational hierarchy—in classrooms, schools, districts and so on.

They recognize the fundamentally diverse nature of their membership and ensure that there is an array of different ways teachers can interact: through caucusing, where teachers with the same interests can develop common proposals and projects, and through democratic decision-making structures, where issues are discussed and debated. They recognize that diverse views and goals are unavoidable and, where possible, attempt to be every teacher's organization.

Rather than countering attacks with the same, they work to challenge negative depictions of teachers, schools, and unions by developing and disseminating powerful alternative ways of communicating their value. For example, rather than falling into the trap of being seen to privilege teachers over children, they will develop ways of articulating the relationship between quality teaching conditions and quality learning conditions. Rather than relying on just a

Vignette 11.1

The following extract illustrates graphically the dangers of union division and the importance of challenging leadership decisions where necessary. It is taken from Javier Corral, "The Politics of Education Reform: Bolstering the Supply and Demand; Overcoming Institutional Blocks" The Education Reform and Management Series *2(1), World Bank, 1999, p.35.*

> Murillo (1999) argues that even more important than union political affiliation are levels of union fragmentation, both internal and external. Internally fragmented unions, i.e., unions whose leadership faces serious internal upheaval, including challenges to the leadership, are likely to contest reforms. When union leaders feel threatened from below, they are more likely to act as "agents of workers." They will feel a greater need to compete for members' votes by challenging state efforts to impose constraints. Union leaders who do not face internal challenges, on the other hand, will feel more comfortable cooperating with the state and even accepting certain sacrifices, as long as there is some compensation.
>
> On the other hand, externally fragmented unions, i.e., those in which multiple unions compete with one another for teacher membership, will be less effective in disrupting reform. In this institutional setting, "each union is weaker, and all of them can only bargain after coordinating their actions" (Murillo 1999, 48). The collective action problems associated with fragmentation reduce the capacity of unions to block the reforms.

slick promotional campaign, they will work on increasing the capacity of educators to learn to promote their own successes and express their own concerns, linking powerful ideas to action.

While most teacher unions struggle under a paradigm of limited resources, strife, and inertia, viable teacher organizations operate with a sense that more is more. Opportunities and ideas are to be exploited; flexibility is key. Recognizing that struggling against the many symptoms of the low value placed on teaching and schooling is an ongoing, long-term necessity, and acknowledging the diversity of teachers' needs and interests, they work on many fronts. A viable teacher union engages intelligently and respectfully with its members, seeking always to be *their* organization. A viable teacher union values democracy in its functioning.

As teachers unions confront escalating assaults on the rights of labor (including the right to bargain collectively), the privatization of public education and the erosion of conditions that support teaching and learning, and the wholesale replacement of one generational cohort of teachers with another, it is critical that these organizations be able to demonstrate that their members drive what they do. Considering organizational vitality in terms of a unions' relationship with its membership is of fundamental importance.

References

Bascia, N. 1994. *Unions in teachers' professional lives: Social, practical, and intellectual concerns.* New York: Teachers College Press.

Bascia, N.1997. Invisible leadership: Teachers' union activity in schools. *Alberta Journal of Educational Research* 43 (2): 151–65.

———. 1998a. Women teachers, union affiliation, and the future of North American teacher unionism. *Teaching and Teacher Education* 14 (5): 551–63.

———. 1998b. Teacher unions and teacher professionalism: Rethinking a familiar dichotomy. In *International Handbook of Teachers and Teaching*, ed. B. Biddle, T. Good, and I. Goodson, 437–58. Dordrecht, Netherlands: Kluwer.

———. 2000. The other side of the equation: Teachers' professional development and the organizational capacity of teacher unions. *Educational Policy* 14 (3): 385–404.

———. 2004. Teacher unions and the teaching workforce: Mismatch or vital contribution? In *Addressing teacher workforce issues effectively: Institutional, political and philosophical barriers*, ed. M. Smylie and D. Miretzky. Yearbook of the National Society for Study of Education (NSSE). Chicago: University of Chicago Press.

———. 2005. Triage or tapestry: Teacher unions' contributions to systemic educational reform. In *International Handbook of Educational Policy*, ed. N. Bascia, A. Datnow, and K. Leithwood, 593–613. Dordrecht, Netherlands: Kluwer.

———. 2008. Learning through struggle: How the Alberta Teachers' Association maintains an even keel. In *Learning through Community: Exploring Participatory Practices* , ed. K. Church, N. Bascia, and E. Shragge. Dordrecht, Netherlands: Springer.

Bascia, N., and N. Jacka 2001. Falling in and filling in: ESL teaching careers in changing times. *International Journal of Educational Change* 2(3): 325–46.

Carlson, D. 1992. *Teachers and crisis: Urban school reform and teachers' work culture.* New York: Routledge Chapman & Hall.

Finley, M. 1984. Teachers and tracking in a comprehensive high school. *Sociology of Education* 57:223–43.

Little, J. W. 1992. Opening the black box of teachers' professional communities. In *The changing contexts of teaching*, ed. A. Lieberman, 158–78. Chicago: University of Chicago Press.

McDonnell, L. M., and A. Pascal 1988. *Teacher unions and educational reform.* Washington, DC: RAND Corporation.

Olson, M. 1965. *The logic of collective action.* Cambridge, MA: Harvard University Press.

Siskin, L. S. 1994. *Realms of knowledge: Academic departments in secondary schools.* London: Falmer.

CHAPTER 12

Challenging Neoliberalism
Education Unions in Australia

Rob Durbridge

School teachers in Australia are part of the much-sought-after middle ground in politics, which major political parties must win in order to govern. They live in "middle class" suburbs of "aspirational voters." The current cliché is "middle Australia," the holy grail that both conservative and erstwhile social-democratic parties seek in order to win office in a political landscape dominated by media imagery and bipartisan neoliberalism.

Yet unlike most of middle-Australia, school teachers are highly unionized and evidently committed to union membership. This applies to both the public and growing private sectors of school education. Private schools are booming under the impact of huge federal government funding subsidies intended to privatize the school system. Yet union adherence continues to grow in that sector, too, where union collective agreements mirror those in the public sector regulating employment.

Teaching is also an aging profession. Around one-third of teachers will retire within the next decade, creating a potential crisis in school staffing. The growth of education union membership has occurred steadily since the current generation of postwar baby boomers entered teaching in the 1970s.

Despite warnings by the unions for many years about the looming teacher shortage and the decline in the attractiveness of teaching as a career, conservative funding policies continue to restrict places in teacher education courses and impose higher costs on gaining entry to tertiary education. Teaching is uncompetitive with other professions after the first eight years or so in terms of remuneration and workload.

In common with other countries where public education has been inadequately resourced, those responsible set out to blame the teachers. Recent research has been paraded by the federal government showing evidence of "poor quality" teachers, who must be "weeded out." It is indisputable—and regrettable—that for some years student teachers were drawn from a lower decile of the student population, which reflects the decline in attractiveness of teaching as a career. This is not helped by a federal government that not only directs most of its school funding disproportionately to the private sector at the expense of public schools but has mounted a political onslaught against public systems and teachers as it has attempted to demonize refugees and other minorities.

Surveys undertaken by the largest education system in Australia, in the state of New South Wales, has found that unlike previous generations of teachers, new teachers do not intend to stay in the profession for a lifetime, in reality a tenure that often turns out to be less than five years. This reflects the changed nature of the workforce as a whole, where aspirations for lifelong careers have been replaced by higher job mobility, as well as the fact that teaching salaries and promotion opportunities are uncompetitive with other comparable professional groups after eight years or so.

The Federal election campaign of 2007 saw United States–style "performance pay" based on student results as a major plank of the conservative coalition's policies, linked to their attacks on collective bargaining and union rights. This was widely seen as a stalking horse for their drive to elevate individual employment contracts over collective instruments and to blame teachers for problems in the education systems, which were highly exaggerated for the purpose. Even the government's own research conducted for it by the Australian Council for Educational Research (ACER) dumped on United States–style performance pay schemes as unworkable.

The conservative government was defeated largely due to rejection of its antiunion laws including its attack on teacher unions. The incoming federal ALP government promised substantial increases in teachers' pay linked to the achievement of teaching standards. The AEU is promoting professional pay and conditions reform linked to performance of professional standards. Collective bargaining will be at the center of the industrial relations system, replacing individual contracts, which will be abolished.

A major challenge for the education unions in the face of the demographic realities is to recruit tens of thousands of new members as their retiring unionized colleagues depart. Thus far, there do not seem to be significant difficulties in achieving this, probably due in part to the politicization of teachers, prompted by the union's activism and membership involvement, in response to the federal government's antiunion reforms. Thousands of new teachers are joining every year, making education, with 260,000 unionized employees, the most unionized industry sector.

Industrial Rights Campaign

Since 2005, Australian education unions have joined with all other unions in a major public campaign to oppose federal antiunion laws. The conservative coalition government has set out to remove the industrial conciliation and arbitration system that has existed for more than a century and that encouraged collective bargaining through unions and recognized the rights of employees. Instead, and in contravention of adopted International Labor Organization (ILO) conventions, the new laws remove many union rights and elevate statutory individual contracts, called Australian Workplace Agreements (AWAs), above both common law and collective bargaining.

Polls show that industrial relations as an issue seems likely to influence the votes of large numbers of voters. A poll of union members conducted by the Australian Council of Trade Unions (ACTU) found that of those who voted for the current federal government in the 2004 election, around 30 percent, most say that they would not do so again due to the industrial relations laws. Changes to the laws were not prominent at the 2004 election and only became so when the conservative coalition gained a majority in the Senate, the federal second chamber. Given that union members alone comprise enough voters to make the difference in a number of marginal seats, the union campaign to convince their members in those seats can be decisive. In some seats AEU members alone can make the difference.

The central union's media and publicity campaign against the federal antiunion laws is unprecedented in scale and cost. Many millions of dollars are being spent on prime TV and radio time, as well as in organization in marginal seat campaigns. The federal government has made some revisions of industrial law in response, but not enough to redress the massive imbalance that was created in the system by antiunion provisions.

Education union members have contributed their share to fund this labor rights campaign, and when decisions have been taken to expend large amounts of members' money, there was no opposition raised against the governing bodies of education unions, nor was there objection from the membership, despite the reality that some vote for the conservative side. Conservative voting members have become conspicuous by their public absence since the antiunion legislation was passed without an electoral mandate.

The union movement's high-profile campaign has included large rallies and demonstrations, national video hookups of union members across the country, and of course thousands of smaller meetings of members in workplaces. The union movement has left behind the political parties in the scope and vigor of its campaign. The political parties now face an extra-parliamentary movement of opposition that is unprecedented in scope and scale. It is the unions that have emerged as the primary voice of opposition when conventional wisdom had it that they were slipping toward extinction and were out of touch with public sentiment for "choice" and "flexibility" in industrial relations.

The education unions have been major contributors to this public campaign, which will have raised more than $10 million for use in mainstream and local media, and campaign events. Major rallies uniting union members and local community members have made events in the campaign occasions for families to participate in activity. Education union members have been prominent in campaign activities, including by taking strike action in some instances.

During 2006 the education unions in all states and territories rallied large numbers of members in public demonstrations and actions. The use of live video teleconference facilities to reach major venues on big screens and hundreds of smaller venues across the country and the outback enabled co-ordinated national union action never before seen in Australia. Tens of thousands of teacher union members participated in actions totaling nearly half-a-million union and community members. At a major action in December 2006, teachers were evidently the biggest group of employees taking part.

Solidarity Action

The involvement of education unions and their members in general union campaigns around industrial rights is by no means novel. In recent years conservative state and federal governments have attacked employment rights, resulting in major disputes requiring—and resulting in—solidarity.

In 1998 the Maritime Union of Australia was involved in a legal assault by the court involving the federal government, waterfront employers, and shady connections with the military and intelligence forces. The purpose of the legal assault was to free the waterfront of unionized labor in favor of labor-hire contracts with workers trained in Dubai. The defeat of this conspiracy was a high point in the union movement's capacity to win public opinion in the face of the power of the national government and the private media corporations, which are generally antiunion.

The education unions were quick to take the significance of this dispute to their members, to illustrate the links that existed between the destruction of the militant waterfront union and subsequent attacks that would occur on unions such as those in education. In addition, considerable funds were raised from education union members and the staffs of education union organizations, which helped to sustain the maritime union members who were on strike or effectively locked out of their workplaces by the employment of scab labor.

During the 1990s conservative governments in both Victoria and Western Australia enacted antiunion laws in the state industrial jurisdictions that had direct effects on members of education unions. The abolition of the award system in Victoria by a radical right wing government, which removed collective industrial rights in favor of individual contracts, energized the union movement to rally in huge numbers in a campaign in which the education unions took a prominent part. The Victorian state government canceled union dues collection arrangements put in place by contract, which effectively reduced membership to zero on the eve of the long Christmas holiday period. The AEU was quick to mobilize to rebuild its membership, involving a massive organizing effort including organizers brought from other states for a short time. The AEU took legal and industrial action in defense of its rights, ending up in the High Court of Australia, where protection by the federal industrial system was achieved.

A similar radical right wing government in Western Australia introduced fines and penalties for legitimate union action. This challenge was met by a determined union movement in which the AEU state associated union played a prominent part. The AEU raised funds from members around the country to assist their colleagues.

In these instances, education unions organized high levels of public opposition to these measures, involving union rallies, strike actions, and extensive public campaigning. These campaigns resulted in the measures—and the governments that introduced them—being defeated. Simultaneously, the education unions involved saw their memberships grow, despite threat of loss of employment rights and conditions.

In Australia we have charted union membership over the past fifteen years. Membership growth parallels activism; when the membership is involved in industrial or political campaigns, recruitment is at its strongest. Or in other words, the greatest enemy of union membership growth is a bureaucratic mode of operation that does not involve members in any real way.

Antiwar Movements

Australian education unions have also been prominent for several decades in social issues and campaigns that go beyond their immediate interests in terms of employment rights and conditions.

One of the most high-profile of these was the unions' opposition to Australia's 1991 involvement in the war against Iraq's occupation of Kuwait and more recently the 2003 U.S., UK, and nominal Australian "alliance" to invade, occupy, and install a compliant administration in Iraq. The education

Vignette 12.1

This excerpt from a World Bank publication explains how the strategy Rob Durbridge outlines in his essay can be undercut. The advice is taken from Javier Corral, "The Politics of Education Reform: Bolstering the Supply and Demand; Overcoming Institutional Blocks," The Education Reform and Management Series *2(1), p. 38.*

> Seeking to isolate a societal player is always politically risky because it fuels the ire of anti-reform players and signals a lack of commitment to participation, which can tarnish the credibility of the reform process. However, when recalcitrant opponents are involved and show no signs of yielding despite the best efforts by change teams to persuade them, working around them might be the only choice. It is crucial at this point to mobilize new coalition partners. This is precisely what the state government of Victoria, Australia, did after 1992. A new reform-minded minister advanced education reforms by avoiding corporatism and deploying instead strategies to counterbalance the coalitional possibilities of unions. The minister avoided confrontations with the unions and instructed bureaucrats to do the same. He even ceased mentioning the unions in public. . . . In short, the government built a strategic alliance with outsider actors as well as one crucial potential cost-bearer (principals), which effectively preempted the coalition possibilities of veto groups. Serious education reforms inevitably produce losers. Whether or not these losers take active stands against the reforms may depend on certain institutional variables: (a) strong links between veto groups and opposition parties in polarized political party systems, (b) the status of executive-legislative relations, (c) leadership challenges inside and outside the unions, and (d) strategic coalitions between veto groups and other societal groups. Of these, (b) and (d) seem to be the most malleable by government policies. These are areas in which governmental policies can overcome institutional blocks.

unions' opposition was among the first and most vocal of community voices prior to and since the wars concerned, regardless of the policies of the major political parties.

While these policies attracted the ire of government and right wing forces in Australia at the time, actually the unions' views have mirrored public opinion that sponsored huge rallies of opposition to the invasion in particular and saw the unions' banners prominent in the streets. In January 2003 the AEU Federal Conference unanimously carried a decision entitled, "No War," which included the following:

[Be it resolved that] as an attack on Iraq will cause widespread death and destruction, increased insecurity and poverty in the world and threaten a wider conflict in the Middle East and elsewhere including the South-East Asian region, the AEU is opposed to war against Iraq. Attacking Iraq to achieve "regime-change" is not only illegal but is likely to be counter-productive and lead to the strengthening of the repressive apparatus of the Iraqi government. Joining an attack simply because the US wants Australian support is inadequate justification and has led to tragedy in the past.

Recourse to war is never justified before all non-violent solutions have been pursued. History is replete with examples of politicians committing other people's children to certain death, trauma and injury in war. The process of UN inspections and deliberations should not be pressured by US bullying of Security Council members into premature decisions. Neither should an Australian decision be pre-empted by a manipulated crisis created by the presence of huge US and British forces in the Gulf.

The Australian parliament must have the opportunity to consult and debate the question of any military commitment prior to any decision. At that time those opposed to a military commitment by Australia will have the opportunity to assess any facts which the government or any other parties bring forward and to put their views before the community and the parliament.

[Be it resolved that] Education International be advised of the views of the AEU on the threat of war in the Middle East and be asked for its support in a campaign to achieve a peaceful resolution of the crisis.

The AEU will participate in and work to build the widest coalition of anti-war forces ever seen in Australian society such that the rush to war in Iraq by the Coalition Government will be rejected overwhelmingly by the Australian community and will mark the Coalition forever as unfit to govern Australia and lead to its removal at the next Federal election.

If the Howard Government commits Australia to an unjustified war, the AEU urges the people of Australia not to support or co-operate with the war effort in any way.

This policy was subsequently endorsed and replicated by many state branch policies. In all states and territories of Australia the AEU then officially sponsored, promoted, and advocated this position in the biggest anti-war demonstrations ever seen in Australian cities and towns. Millions joined rallies of opposition. The Australian government, locked into the U.S. strategy as it was, ignored public opinion, which ran three to one against involvement, and committed a military force that was token in effect, other than to give political credence to the so-called Coalition of the Willing.

It reflects the expectations that educators have of their unions in Australia that this high-profile campaign attracted only the most minimal criticism from individual members and was overwhelmingly supported by the rank-and-file membership.

International Solidarity

International policies have often attracted the interest and involvement of education unions in Australia. The campaign for the independence of East Timor is among the highest profile and longest running of these. Perhaps ironically, when the Indonesian-sponsored militia violence occurred after the success of the UN-conducted referendum for independence in 1999, the education unions joined a wide spectrum of religious and community opinion in calling for Australian military action to defend the people of East Timor. In this instance, the unions were calling for peacekeeping troops to restore order in a neighboring country.

On the domestic front, and related to education more directly, the education unions took a prominent stance against the Conservative government's post 9/11 scare campaign to demonize refugees as threats to peace and security. This involved the creation of prison camps in remote locations in the outback and on impoverished Pacific islands to locate refugees who had attempted to "illegally" enter Australia.

So-called detention centers were established to "process" refugees under the government's chauvinistic slogan, "We will determine who comes to this country and the circumstances under which they do so." These centers are administered by private contractors such as the Wackenhut Corporation and Group 4, which are corporations running prisons in the United States and Britain, and they are often located in remote and hostile parts of the country, such as Woomera. Refugees, who were often traumatized and desperate when they were taken there, were mistreated and shamefully abused.

Education unions, prominent from the beginning in opposition to these policies, issued media statements, took part in protests, submitted parliamentary enquiries, and defended parents and children denied access to normal schooling. The AEU organized among schools in the localities of camps to offer places to children so they and their parents could leave the centers to attend schools.

The unions focused on the incarceration of children in refugee camps and the denial of their right to education in contravention of the UN Declaration of the Rights of the Child. While the Labor opposition was initially stampeded into support for these policies in the belief that they had public approval, a determined campaign by civil libertarians and religious groups succeeded in winning public opinion to oppose the harshest aspects of these policies.

Labor then adopted the union's views on the release of children and their parents to live in the community pending the determination of their cases, a view that the government ultimately adopted. Subsequently the government has rehoused most families out of detention and allowed their children to attend schools, while some detention centers have been closed.

International Aid and Training Programs

The Australian Education Union has conducted a major international program of aid and assistance to education unions in the geographic region in one form or another for over thirty years, funded entirely from union membership dues. The program was inspired by the work of the Canadian Teachers Federation (CTF) in the 1960s, supported by the Canadian government at that time. In the mid-1970s the Papua-New Guinea Teachers Association and the Indonesian Teachers Union received training and physical infrastructure assistance at difficult times in their histories.

This national initiative had been prefigured by efforts at regional assistance provided by some state teachers unions, particularly in New South Wales and Victoria.

The International Trust Fund (ITF), established by the AEU in 1981, was established as formally independent of the governance of the union but elected by the same constituency that elects the officers and executive of the union. By this means the fund has been able to qualify to attract Australian government aid funds for some of its projects, although this has ceased under the rule of conservative governments for a decade or more.

The ITF, usually in conjunction with the international organization Education International and its predecessor organization, has assisted teacher organizations in Central and Latin America and the Chilean Teachers Union for its rebuilding following its suppression by the Chilean junta.

Funds have been provided for a broad range of social justice and economic concerns, including support of teachers in Vanuatu who had been dismissed for taking strike action, of the Alliance of Concerned Teachers in the Philippines, of the Kanak union in New Caledonia to develop an independent press and media, of an English as Second Language teacher for the All-China Federation of Trade Unions, of the Papua-New Guinea Teachers Association for membership development, and of the Indonesian union for a training program for women members.

Since 1986 the ITF has contributed with Nordic and Netherlands unions to "unity building" among unions belonging to the black African Teachers Association of South Africa, which led to the the creation of the South African Democratic Teachers Union (SADTU) in 1990. From being 80 percent dependent on these funds, involving in total around U.S.$250,000, SADTU was able to become independent of the consortium and achieve a key role in the campaign against apartheid and the building of a new education system in a new country.

In 1995 the ITF clarified its purposes to emphasize the promotion of core human and labor rights on a universal basis. The decision of the ITF stated its intention to "promote the development of united, independent and strong education unions through training courses and other effective means; develop women's participation and leadership in education unions; provide solidarity assistance at times when a union may be facing oppression; provide emergency/humanitarian assistance in times of crisis; promote the principles of human and labor rights amongst the leadership and membership of education unions; pursue important humanitarian issues such as child labor and campaigns against illiteracy and the oppression of women, indigenous peoples, and other sections of society" (ITF 1995, 3).

Union Policy and Action on Equity, Discrimination, and Racism

The education unions in Australia take policy decisions on social inequity, discrimination against women in employment, and discrimination on the grounds of sexuality or sexual preference, as well as racism in all its forms.

However, in addition to making policy decisions, the union puts these priorities into practice as well. The AEU has employed Aboriginal Education Officers at federal and many state levels since the 1980s to work on issues affecting indigenous students and teachers, as well as indigenous education and reconciliation in society generally. Women's Officers are employed by most unions to address the inequality that women face in gaining promotion in education systems and discrimination in various forms.

Through the ACTU the education unions pursue the social policy objectives in seeking reform of legislation and in campaigns to win public support.

Education Unionism in Australia

The membership of education unions now comprise Australia's biggest single unionized occupational group, which also has the union movement's highest rate of union density. The three Australian affiliates of Education International that are also affiliated to the national center, the ACTU, together have a membership of more than 220,000. There are no other education organizations other than a small union of school principals in two states. The school education unions are the Australian Education Union and the Independent Education Union, organized respectively in public and private schools, and the National Tertiary Education Union, organized in universities.

Most school teachers and allied educators are unionized, with more than 80 percent union density, matched only by that in nursing, a comparable professional area. Unionization rates of school teachers in both public and private sectors are higher than those in any other blue- or white-collar area of employment other than in concentrated areas such as major project construction and the waterfront. While education unions continue to grow, this comparison is also a measure of deunionization in the private sector because a decade ago the education unions were smaller in total than a number of both blue- and white-collar private-sector unions. The education unions together have more members than any other single union. Technological change, production shifts to offshore locations, and antiunion human resource management have reduced traditional unionized sectors, and new sectors such as information technology remain largely ununionized.

In contrast to a general decline in unionization since the 1970s, the education unions have built their memberships in both the public and private sectors. School privatization policies in the form of high subsidies paid to church-operated schools have concentrated growth in the private sector. This has been reflected in higher growth of private-sector union membership at a level of around 65 percent. Around one-third of school teachers work in private schools, reflecting the enrollment share attracted by subsidies that subtract from and undermine public provision. Successive governments at state and national levels have followed this neoliberal policy, notwithstanding their party nomenclatures.

A survey conducted in 1999 by the principal union in the sector, the Australian Education Union, called "Building the Foundations," found that women members gave collectivism and solidarity as the main reason for joining the union. Salaries and working conditions were ranked second behind ideological reasons to do with the values of unionism and the collective enterprise of teaching.

Like the factories that gave rise to the labor movement's big battalions in the nineteenth century, school (and health) systems comprise large concentrations of workers whose employment in Australia is still expressed through collective arrangements; union collective bargaining is the norm. While unionism among public school teachers was established in the 1890s, and despite a spectacular and long strike in Western Australia in the 1920s, unionization and industrial strike action did not become widespread until the 1970s. This reflected the times; growing enrollments caused staff shortages, necessitating recruitment of many young teaching staff from radicalized university campuses.

The employment and industrial relations arrangements of teachers in Australia reinforce a collective outlook. They are generally treated by the state and most private employers as part of a single "enterprise" in which single union collective agreements are negotiated. The education unions have progressively gained increased membership among nonteaching school staff, with shared union coverage being the norm in most states.

Devolution of educational administration has occurred to some extent in some state systems, but competition for salaries and positions has until now been relatively restrained in systems that transfer staff and negotiate union collective agreements for an entire state or systemic workforce.

Due to the hostility that both state Australian Labor Party (ALP) governments and Catholic school employers have adopted to the federal coalition's current aggressive antiunion policies, known as WorkChoices, collective bargaining is likely to remain the norm in both public and private sectors, with pay and conditions outcomes running closely in parallel across the sectors. Of course, these policies could change, particularly due to the strong hand the federal government can play as the primary source of funding for private schools and as a significant contributor to state public school funding.

Individual Employment Contracts

The election of the federal ALP government at the same time that state governments are in ALP hands creates unprecedented opportunities for improvement to public education. In place of the threat by the former government to bypass state employment agreements to introduce performance pay, the new government will introduce a national industrial relations system based on collective bargaining. The new government can replace coercive federalism with the cooperation between employers and unions needed to improve public education and teachers' working conditions.

The incoming federal ALP minister for education, Julia Gillard, who is also deputy prime minister and minister for industrial relations, was a product of and a strong advocate of public education in the political opposition. The unions expect the new government will work collaboratively with state government administrations and unions to implement reforms rather than attempt to override them and impose individual employment contracts involving performance pay, as was the former minister's plan.

The unions face the problem that the ALP has become captive to neoliberal orthodoxy. At state and federal levels, it has endorsed reducing funding to the public sector while promoting the private sector, thus producing one of the lowest levels of expenditure in the OECD. While the antiunion policy of the previous federal government will be reined in, funding to redress inadequate conditions in the public sector is a battle to be won. The education unions are a natural constituency for defense of adequate funding as teachers put quality education first.

The resistance of the unions to these policies has caused successive governments, regardless of their political brand, who function in the public sector as employers, funders, and regulators of education, to adopt a wary dislike and even hatred of education unions. Bipartisan neoliberal budget policies are often at odds with union campaigns that link to community support for quality provision and better pay and conditions for educators. Classic education union campaigns focus on reducing class sizes and improving the quality of school infrastructure. Conservative and sometimes Labor governments in a number of states have mounted head-on attacks on the public education unions in an attempt to break union support over recent years, but despite setbacks, the unions have been able to rebuild membership. In the largest state, New South Wales, a Labor Government attempted to bypass the union by taking industrial bargaining directly to teachers using the departmental e-mail network. Graphic images of a teachers' demonstration hurling copies of the government "offer" across the fence of Parliament House was the result.

ALP governments in office in all states have promised much and restored the worst cuts imposed by conservatives, but they then have often pursued generic neoliberal budget policies, which has resulted in the erosion of the ALP's traditional support from those who rate public education as an important issue and of teachers' allegiance to the ALP in particular. Some voters have left Labor for the conservative side, but more direct votes have gone to the Green and alternative independent candidates, which in Australia's preferential electoral system come back to the ALP.

Government cuts to education funding (usually under the guise of introducing a "new deal" of one kind or another) have repeatedly foundered when the union has been able to mount an effective campaign of resistance that wins community support. Despite this, the real incomes of teachers have fallen over several decades along with those of most groups of employees other than managerial strata.

Much of the above is familiar to education unionists in the Organisation for Economic Co-operation and Development (OECD) and developing countries. One of the striking experiences of education unionists who meet their comrades at international conferences is the similarity of the policies being pursued across the globe.

Defeating Corporate Globalization in Education

Teachers and education workers are at the forefront of the clash between community expectations of their governments and the role that corporate globalization wants governments to play. On the one hand, the community wants services that improve the life opportunities of its children from one generation to the next. On the other, the corporations want more access to public funds for their own purposes and garnish their pursuit with ideological values such as competition and devolution as if they were self-evident truths.

Education unions attract the anger and revenge of governments precisely because they have credibility in the community, due to their members' daily contact with millions of students and their families in communities in which they themselves live and work. AEU polling shows that teachers and their unions speaking on educational issues are given far more credence than politicians because of our professional expertise and undeniable commitment to the best educational outcomes.

As is shown by the AEU's experiences, campaigns by education unions and the mutual support of campaigns by and with similar organizations in the community can build alliances that can win public opinion and turn this into a powerful electoral factor. This requires consistency, good research, sophisticated use of the media, and above all independence and genuineness. Unions that are a part of political parties or factions cannot achieve the same level of public credibility. The AEU's polling and focus group research shows that public agreement with the union's policies and objectives is high compared to that of political parties and governments, which are seen to have agendas of their own rather than the interests of education at heart.

While the political left has been largely checkmated by the success of neoliberal policies and ideology, the activism of the community and union-based movements of opposition and change has never been higher. Participation in rallies against Australian involvement in the Iraq war were far bigger than anything seen during the campaign against the war in Vietnam, for example, and there is far more local involvement in local environmental campaigns than ever before.

The last bridge that must be crossed is for unions to bring those movements into the political center to realize their aspirations backed by government action. That would mean the defeat of corporate globalization's aspiration to mold the world in its own image.

PART IV

Teaching, a Profession under Attack

CHAPTER 13

Contradictions and Tensions in the Place of Teachers in Educational Reform

Reflections upon the Role of Teachers in Recent Educational Reforms in the United States and Namibia[1]

Ken Zeichner

In this essay, I draw on my own experiences over the last thirty-seven years as a primary school teacher and teacher educator in the United States, and my experiences from 1994–2004 participating in and then studying the educational reforms in Namibia. Specifically, I compare educational reforms in the United States and Namibia to examine two dichotomies that have been a central part of debates about teacher quality throughout the world (teachers as technicians vs. teachers as reflective practitioners and teacher-centered instruction vs. learner-centered instruction). I will argue for a way of resolving the tensions associated with these binaries that addresses the legitimate interests of both states and teachers in shaping educational reforms, one that takes account of recent research about teacher and pupil learning and the limitations found in teachers' working conditions throughout the world.

Teachers as Technicians vs. Teachers as Reflective Professionals

Although everyone agrees that teachers are the most critical element in determining the quality of a nation's educational system, there is much disagreement about the role that they should be prepared to assume in schooling and the preparation and support that societies should provide to them. At one extreme is the focus on preparing teachers at low cost to be low-level technicians and civil servants who obediently follow a scripted curriculum and prescribed teaching methods. In the United States a high official in the national education department has referred to the preparation of teachers who are "good enough"—just good enough to follow a scripted curriculum and be trained in prescribed teaching practices, allegedly based on research. He and others in the Bush administration have claimed that tightly monitoring teachers' actions, scripting the curriculum, and intensifying standardized achievement testing with serious consequences for schools and teachers related to the examination results will lead to a raising of educational quality and a narrowing of the achievement gaps between different groups.

Apart from the argument that standardized achievement tests measure only a very small part of what the public wants from its schools, the national media and a number of scholars in the United States exposed the fraud involved in the improvement of Texas schools under Bush policies, the so-called Texas miracle (Valenzuela 2005). Advocates of preparing teacher technicians argue that many children in U.S. public schools, particularly poor children and children of color, have less access to fully qualified teachers who have completed a teacher education program and that these "good enough" teachers who are trained to follow directions but not to think and exercise their judgment are better than the teachers who are just pulled off the street into classrooms with no preparation at all.

This same argument has been made in many developing countries that struggle to give all students access to teachers who have completed a full-scale teacher education program at the postsecondary level. With increased access to basic education, and in some countries to secondary education, and with the implementation of neoliberal economic policies that have resulted in drastic reductions in public expenditures in many countries (Carnoy 2000; Reimers 1994), it has become increasingly difficult to provide fully qualified teachers for every child.

Some say that providing a fully qualified teacher to all learners around the world is an unrealistic goal unless we move to more "cost effective" training of teacher technicians because of the limitations of teachers' working conditions in developing countries and because of cultural factors. Consequently, another part of this efficiency strategy is to develop more "fast track" preparation programs to get people into classrooms more quickly, often with very little preparation before they become responsible for classrooms.

I find it interesting that many of these government officials who advocate the "good enough" teachers do not apparently find these teachers good enough for their own children, whom they often send to private schools. I would have considerably more respect for the policies of the people who advocate cost-effective teacher preparation (teacher education on the cheap) if they were sending their own children to be taught by these low-cost teachers.

In contrast to preparing teachers to do but not to think and exercise judgment are efforts to prepare teachers as reflective professionals who are given some discretion to exercise judgment at the classroom level about how to adapt the curriculum and instructional methods to best teach the students they are responsible for educating—or as Samuel (2005) has described, to make situated and interactive judgments appropriate to contexts (also see Swarts 2001). The curriculum may be provided at a national level and some support given to teachers with regard to teaching strategies, but there is also commitment to actively involving teachers in interpreting the reforms and adapting them to meet the diverse needs of their learners, to gaining teachers' commitment to both the underlying ideas behind the reforms and the practices associated with them, and with providing them with professional development opportunities that support their implementation of the reforms and their adaptation to varied circumstances.

Much of the history of educational reform around the world falls somewhere between these two poles. The preparation of teachers as reflective professionals is not a panacea for the problems in public education around the world. One thing that I have learned over the years as a university teacher educator is the importance of providing support, scaffolding, and direction for prospective teachers to gradually develop over time the knowledge, skills, and dispositions needed for teachers to play an active role at the classroom and school levels in exercising their judgment about matters of curriculum and instruction, and in participating with others, including parents, in the making of policies affecting schools. Sending student teachers out to schools with little guidance and support and merely telling them to reflect can be just as much a disservice to teachers and their pupils as efforts to train them to mindlessly carry out teaching scripts.

The preparation of teachers as reflective professionals has been difficult to achieve in most countries in part because of severe cuts in public expenditures for services, including education, and the costs in budgets for teacher salaries. The sheer size of the teaching force worldwide, over 60 million teachers (UNESCO 1998), has made it difficult for most countries, including my own, to put fully qualified teachers, those who have completed secondary education and some postsecondary preparation for teaching, into every classroom to teach subjects for which they have been prepared. Investing in the preparation of reflective professionals who are provided with the working conditions (salary, physical facilities, instructional materials, class sizes, opportunities for continued learning, and so on) that are consistent with an elevated professional status creates a further challenge. These conditions create a tension over the degree of autonomy given to teachers so that they are given direction and support to learn and implement particular teaching strategies that lead to enhanced student learning.

Outside-Inside, Top-Down Educational Reform

Despite the various ways countries have worked out the tension between preparing teachers as technicians or reflective professionals, and despite the worldwide rhetoric about making teachers reflective practitioners, many educational development projects have employed an outside-inside, top-down model that emphasizes the preparation of teachers to obediently carry out a plan conceived and developed by others removed from the classroom. This occurs in part because of the widespread influence of development agencies such as the World Bank. Samoff (1999) has observed:

> The annals of educational reform, especially in the "third world," are replete with efforts to reduce the role of teachers or replace them with something else. . . . Especially where a large percentage of the teachers lack the required formal education, strategies for improving education as a delivery system often involve efforts to regiment or alternatively bypass the teachers. The combination of restrictive teachers' guides and curriculum manuals, the lack of instructional materials, and the consequences of national examinations leave teachers very little role in determining what is to be taught and how to teach it. . . . Except in the most minor ways, they are not expected to be creators, or authors, or inventors, or imaginers, or even learners (p. 20).

Villegas-Reimers and Reimers (1996) describe a failed national educational reform in Pakistan that typifies the dominant approach to teachers by policy makers in many countries.

> In Pakistan in the early 1970s, the Ministry of Education spent several years designing an innovative approach to improve quality in primary schools. They came up with a teaching kit that included about 100 items such as beakers, a national flag, an abacus, and other materials. 60,000 of these kits were produced and delivered. The kits were based on a pedagogy that expected student participation in small groups and learning from direct experiences rather than using the traditional pedagogy (based on rote memorization) in which teachers

were trained in Pakistan. In a survey conducted in Pakistan by one of us, we found that few (about 1 in 5) of the teachers used it. By "using it" these teachers meant using it an average of 7 times during the school year. (p. 474)

Notice that this report on the effects of the reform does not address the quality of the use of the materials or how their use affected the learning of pupils. Even with the lowest possible standard, use of any kind, the results are quite meager.

This situation of reformers ignoring the learning needs of teachers has been repeated over and over again in countries throughout the world. My first years as a primary school teacher in New York came at a time when eminent scientists in the United States with millions of dollars of funding from the government and NGOs had just completed the development of "teacher-proof" science kits that were to transform science teaching throughout the United States. Studies have shown that most of these kits sat in school closets and were used very infrequently by teachers who often had little understanding of the new pedagogical ideas underlying the materials and limited knowledge of the science content. The reformers, experts in science content knowledge and some of the leading scientists of the country, had decided to ignore teachers and tried to bypass them by developing tightly regulated scripts for science teaching that teachers were to mindlessly carry out. Understandably, teachers largely ignored these scripts and went about their science teaching largely as before.

Despite the efforts of policy makers to provoke all kinds of changes in classroom practice, in the end, teachers have been very successful in subverting these efforts "behind the classroom door" (Goodlad and Klein 1970). Announcing changes in schooling, even demanding these, will not change what happens in schools and classrooms if teachers do not implement the changes.

It remains to be seen whether the new mechanisms of surveillance and control through standardized testing and putative accountability systems based on these assessments will change things. The evidence so far in the United States is that they will for some students, but not in ways that support one of the central goals of public education in democratic societies, to provide a high-quality education to all children that enables them to actively participate as citizens in the political and economic life of their societies and to lead decent and rewarding lives (Barber 1997). The punitive consequences associated with standardized tests are steering instruction toward a narrow focus on reading and math skills while ignoring elements the public in many countries wants from its schools: higher-order thinking and problem-solving skills, moral development and development of the ability to get along with one another and to resolve conflict peacefully instead of through violence, aesthetic learning, and civic learning that enables students to become active participants in democratic societies (Dillon 2006; McNeil 2000).

A New Focus on Educational Quality and Learner-Centered Education

One aspect of this expanded definition of Education for All has been transforming teaching and learning. Specifically, in many countries efforts have been made to move away from autocratic teacher-centered classrooms that focus on rote repetition of reified content (that is often unrelated to students' life experiences and denies their cultural and linguistic backgrounds) to teaching that is more learner- centered and culturally relevant.

Although definitions of learner-centered education vary among and within countries, certain features are shared. These include valuing learners' life experiences and current understandings as a starting point for instruction, respecting the cultural and linguistic resources that pupils bring to schools instead of viewing them as deficits if they are different than the dominant ones, using local materials and natural resources as part of the curriculum, and moving away from an overreliance on commercially produced curriculum materials; fostering a higher degree of learner participation, discussions, and contributions within classrooms; focusing on learner understanding of subject matter and not just on memorization and rote repetition of isolated facts; and focusing on developing learners' abilities to use knowledge acquired in school in meaningful and authentic life situations. In addition, in some countries there is an increased emphasis at all levels on democratic involvement in the educational decision-making process.

One perspective in the literature on educational development suggests learner-centered education is beyond the capacity of teachers in developing countries, where the working conditions for teachers are more limited than in much of the developed world. Scholars have also sometimes argued that these models have acted as colonizing pedagogies that lead to the destruction of traditional modes of thought while facilitating the penetration of capitalist ideology in "periphery states"(Guthrie 1990; Tabulawa 2003). Despite these criticisms, evidence from studies of teacher education in Namibia illustrate that learner-centered teaching is possible and useful in the global south.

Teacher Education and Educational Reform in the United States

Many teacher educators in both the United States and Namibia have sought to stimulate a shift toward more learner-centered and culturally responsive practice in classrooms. This has been the official policy of the Namibian government since 1990. On the other hand, U.S. government policies support scripted classroom practice and a diminished role of teachers in schooling. The Bush administration has engaged in a systematic attack on programs of teacher education within higher education to deregulate teacher education and marginalize universities' roles in teacher preparation and professional development (Zeichner 2003). One tactic to shake public confidence in teacher education has been to equate the commitment to learner-centeredness and multiculturalism that characterize many programs of teacher preparation in the United States with a lowering of academic standards (e.g., Walsh 2004).

The attack on teacher education in the United States has to be understood in reference to the fundamental crisis of inequality in its public schools discussed by Lois Weiner in her essay. Despite two decades of progress in narrowing the achievement gaps between poor and minority students and other students, this progress has stopped, and the gaps have begun to widen again (Haycock 2005). One aspect of the crisis is the failure to provide even minimal levels of quality to much of the school population that is working class, minority, and poor, as measured by such things as attendance, school dropout rates, and performance on standardized achievement tests (Lipman 1998). Taking into account other educational outcomes not measured by standardized tests, such as higher-order problem solving and reasoning abilities, the picture is even bleaker (Peske and Haycock 2006).

Those who defend these standardized achievement examinations admit that there are important outcomes that they do not address but say that these are the best measures that we currently have available for measuring the success of educational systems. What they argue is another way of saying that these tests are "good enough" to assess the work of the "good enough" teachers that they hope to prepare to teach other people's children.

Bush administration efforts to limit teacher education to training in using scripted curricula is proposed as "scientifically based" but is, in reality, ideologically driven, (e.g., Allington 2002). The rapid growth of alternative teacher education programs in the United States (Baines 2006) makes the United States fit with the traditional behavioral skill training model of educational development common throughout the world.

Teacher Education and Educational Reform in Namibia

Since independence in 1990, the government of Namibia has taken a different route—until very recently: "When we were in school, we were not given a chance to talk. The teachers used to give us only the summary, life history and social studies. They put a lot of things on the board and we did not understand what was going on. Now we give learners a chance to come up with different ideas so you can see if they really understand" (BETD graduate interview, June 13, 2003, cited in Zeichner and Luecke 2004).

This quote typifies the essence of post independence educational reforms in Namibia, reforms John Nyambe explains are under attack because of pressures from international development agencies such as the World Bank. After independence in March 1990 the government of Namibia attempted to reform its educational system from a very autocratic one emphasizing rote repetition of received knowledge to a more learner-centered and democratic one that focuses on the development of student understanding and that is relevant to and respectful of different cultural traditions and communities (Angula and Grant-Lewis 1997). This new system was to provide a high-quality education to all Namibians rather than only to an elite few, as had been the case for many years of colonial rule and apartheid.

These reforms viewed teachers as agents rather than objects in the reform process. One example was the development of a new three-year national teacher preparation program (the BETD) based in the colleges of education, which emphasizes preparing teachers as reflective practitioners who are able to actively participate in working out the details and necessary adaptations that are needed to give meaning to the broad principles guiding the reform.

In the next section I compare two different sources of data about the impact of these reforms. One is a study that I worked on with U.S. and Namibian colleagues in which we studied graduates of the postindependence teacher preparation program for basic education (Zeichner 2005; Zeichner and Luecke 2004). The World Bank also completed a study of the state of Namibia's educational system. (Marope 2005) and has helped Namibia develop a fifteen-year, U.S.$1 billion–plus plan for reforming its educational system based on this analysis, the Strategic Plan for the Improvement of the Education and Training Sector (ETSIP; World Bank 2007).

While the World Bank acknowledges some of the gains that have been made in Namibian education in the fifteen years since independence (e.g., in improvements in school attendance rates, in growth in the proportion of qualified primary and secondary teachers) and the high commitment of the Namibian government to education as evidenced by the share of its GDP that is allocated to education, much of the 2005 report focuses on problems in the progress of the reform and social and economic factors such as poverty and HIV and AIDS that have influenced the progress of the reforms.

For example, while acknowledging that between 1995 and 2001 the progress that was made in providing qualified teachers to primary and secondary schools of Namibia, the report notes that currently 60 percent of primary teachers and

27 percent of secondary teachers are still unqualified, qualified being defined as having completed secondary education and postsecondary education for teaching of at least 3 years.

In terms of the quality of learners' academic performance, a number of studies are cited in support of the claim that there is still low functional literacy among primary and junior secondary school graduates. For example, it is asserted that in 2001 only 46 percent of the candidates for the junior secondary school examinations (in grade ten) attained the minimum level required for entry into grade eleven. They also cite a UNESCO-sponsored study that concluded that two-thirds of grade six Namibian learners could not read with any level of proficiency. Some attention is also given to patterns in the distribution of qualified teachers and in learner outcomes that disfavor the previously disadvantaged northern regions. The report also argues that although the reform policies endorse a wide repertoire of progressive teaching methods, traditional teacher-centered methods are still common.

While the World Bank study found relatively little to commend, the other investigation, conducted between 2002 and 2004 by U.S. researchers, including myself, joining with Namibian researchers from the National Institute for Educational Development, identified promising developments. We surveyed over 400 graduates of the BETD program that emphasized learner-centered and democratic education who were teaching in different regions of the country and were from different cohorts of the program to assess their understandings of and degree of support for the reforms. We also observed teacher education classes in two of the colleges that offer the BETD program to examine teacher educators' understanding of and support for the reforms and the relationship between the pedagogy of teacher education in the colleges and the principles and ideas underlying the reforms. Finally, we interviewed and observed 74 teachers from different cohorts of the program who were teaching in both urban and rural areas surrounding the two largest colleges of education and interviewed their principals (Zeichner 2005; Zeichner and Luecke 2004).

While I do not dispute the accuracy of the data set forth in the World Bank report on the status of Namibian education, I can say that on the basis of our limited study of BETD graduates that another important phenomenon has occurred. Even under sometimes very difficult circumstances (e.g., large class sizes, limited physical settings and instructional materials), we found not only a great deal of support for the learner-centered education and democratic pedagogy advocated by policy makers among teachers and principals, we also found significant implementation of the key aspects of LCE in the classrooms of teachers we observed and interviewed. For example, our observations revealed widespread use of small group work, classroom discussions, drawing on learners' knowledge outside of school, and teachers' efforts to connect instruction in the classroom to their learners' prior knowledge and experience. We saw less evidence, however, of some of the elements of the reform such as individualizing instruction for different learners and cross-curricular integration.

What we observed in the classrooms of BETD graduates was very similar to the teaching described by O'Sullivan (2004) in her study of English Language teaching among unqualified and under-qualified teachers in remote rural schools in northern Namibia following professional development activities on LCE. O'Sullivan describes an adaptive version of LCE that she calls "learning-centered education," a hybrid of teacher-centered and learner-centered methods that takes into account teachers' working conditions.

This adaptive version of LCE is not the rote and formalized teaching that was common in Namibia before independence, nor is it the idealized version of LCE that is common in policy documents and in caricatures by critics of LCE. This hybrid model of teaching represents a clear break with the past. Our data and O'Sullivan's data demonstrate that it is possible, even under conditions of limited resources, for teachers to learn how to teach in more LC ways and to enact this kind of teaching in their classrooms.

Our study should raise a caution about abandoning the path taken by Namibian educators since independence. Because BETD teachers still comprise a very small percentage of the approximately 18,000 teachers in Namibian schools, one would not yet expect to see the results of these changes in teaching in national outcome data. Poverty, HIV and AIDS, and a host of other factors have also obstructed the realization of the goals for education set in 1990 (Pomuti et al. 2006).

While some of the recommendations made in the World Bank report for dealing with the situation in Namibia make good sense in my view, particularly expanding preprimary and postbasic education and eradicating inequalities in education among different groups, other proposals are worrisome, such as raising class sizes (allegedly for greater cost effectiveness) and "better managing" teacher salaries and focusing on "enhanced lines of accountability" while ignoring genuine professional development for teachers and teacher educators. These prescriptions undermine the core principles that have guided the Namibian reforms since independence. Nowhere in the World Bank documents is there mention of learner-centered and democratic education, the basis for the post independence reforms. In fact, nowhere is there mention of pedagogy of any kind. Instead, the focus is on increasing accountability and performance assessment with no attention to the kind of professional development for teachers and teacher educators that we know is essential for the success of educational reforms (Schwille and Dembele 2007; Villegas-Reimers 2003).

The World Bank recommends that the government establish "quantifiable learning targets" specifying the knowledge to be acquired by learners at every level of schooling based on "formal, externally moderated, and internationally comparable" assessment systems that can be universally applied. The teacher education curriculum, which will also be performance-based, will be aligned with the new standards-based school curriculum, and teachers will be trained to implement it.

While there is nothing wrong per se with establishing a clear plan for what knowledge learners should acquire in the different phases of schooling and monitoring their performance, this type of agenda taken too far easily leads to the very kind of nonsense I described earlier that the Bush administration is promoting. Taken too far, this proposal will lead to an exclusive focus on learner performance, on narrow assessments, to the allocation of scarce resources to the support of an elaborate assessment system, and to viewing teachers as mere conduits who follow scripts provided to them to raise test scores. Other important goals such as the role of education in supporting the development of a democratic society through the critical thinking and problem-solving skills developed among learners will get lost in the process (Meier et.al., 2004).

Additionally, while expanding access to preprimary and postbasic education are important goals, to propose to pay for this by raising class size and limiting teachers' salaries is the wrong approach. The 2007 World Bank document outlining the elements of ETSIP argues that increasing "the efficiency of teacher utilization" by raising class sizes (primary classes to forty to one and secondary to thirty-five to one) and decreasing the proportion of the Ministry of Education budget used for teacher salaries (from 96 percent to 85 percent in two years), together with other rationalization measures and loans, will provide the money needed to implement the standards-based reforms. Ample research demonstrates that limiting teacher salaries will drive teachers from the profession and increasing class size will lower rather than raise educational quality (e.g., Avalos 2000; Grissmer 1999; Ingersoll 2003; Jansen 1999).

Conclusion

What are the implications of these efforts at educational reform in Namibia and the United States for the rest of the world, and how do these experiences relate to the kind of teachers we should be seeking to prepare? Anthony Burke

(1996) has argued that "the decision regarding the future role of the teacher as technician or professional is perhaps the most important one to be made in education today (p. x)."

There is a choice to be made about the direction to go in improving educational quality. On the one hand, we can accept the current place of education in societies, doing the best that we can with what is available to us, focusing on preparing teacher technicians at low cost who it is hoped will obediently carry out the master plans designed by policy makers to raise examination results on a narrow range of basic skills. We have many examples of programs that have cost-effectively staffed classrooms with technician teachers who are able to successfully follow scripts provided to them. This model, while providing "qualified" teachers to schools, serves to perpetuate teaching as a didactic process of one-way transfer of knowledge from teachers to learners, the kind of teaching that has failed to serve the vast majority of learners in many countries. It is clear that we can get teachers to follow scripts and that test scores for some learners will rise as a result. This should not be confused, though, with providing a high quality education.

Another approach is to acknowledge all of the limitations and problems that surround education and to pursue the goal of providing an education to everyone's children that will enable them to participate more fully in both the political and economic life of their countries. While acknowledging the importance of education's role in relation to the economy, we must recognize the danger of only focusing on the economic implications of educational policies (Zeichner 1999). Namibia would come closer to accomplishing its goals if it invested a portion of the money now being put into increased assessment and accountability into continuing and strengthening the professional development for teachers and teacher educators begun at independence.

For several reasons Namibia should seek to maintain its goals to enact a more learner-centered and democratic approach to teaching in its schools— rejecting the hyper-rationalized approach advocated by the World Bank, driven primarily by a narrowly economic view of effectiveness. First, very limited aspects of learning are addressed by the assessments that need to be implemented as a condition for accepting a loan.

The second reason Namibia needs to maintain its course is related to what we know about successful teaching and the process of learning to teach. A report released in 1998 and sponsored by a number of organizations including the World Bank, "Teacher Development, Making an Impact" (Craig et.al. 1998), documents successful examples of teacher education respecting the abilities of even the poorest and most untrained teachers in places like Guatemala and Colombia . The researchers concluded that when teachers are actively involved and empowered in the reform of their own schools, curriculum, pedagogy, and classrooms, even those with minimal levels of education and training are capable of dramatically changing their teaching behavior, the classroom environment, and improving the achievement of their students. Conversely, when teachers are ignored or when reforms come from above, unconnected to daily realities of the classroom and local environment, even the most expensive and well-designed interventions are almost sure to fail. Poorly educated, underpaid, overworked teachers can become reflective, empowered professionals.

Looking beyond Schools

Problems of educational quality and inequity cannot be solved through educational interventions alone. What we rightly expect our schools to do, to educate everybody's children to a high standard, is out of proportion to what governments have been willing or able to invest to make this happen. As a recent UNESCO report on the state of the world's teachers concluded, if education is expected to help the poor lift themselves out of poverty, then education itself must be lifted out of poverty (UNESCO 1998).

While it is important to do the best possible job that we can with the resources at hand, it is equally important that we speak out against the inequitable distribution of resources among and within countries and the ways in which resources are currently allocated to different societal needs.

There is no shortage of money in the world to deal not only with all of our educational problems but with the broader problems of poverty that face millions of people across the globe. For example, since March 2003, when George Bush and his government invaded Iraq, the U.S. Congress has spent U.S.$450 billion to finance the war and occupation, and Bush continues to ask Congress for even more money for the war while proposing cuts in things like early childhood education, food assistance to the elderly, and children living in poverty (National Priorities Project 2007).

The widespread adoption of neoliberal economic policies in both developed and developing countries has resulted in major cuts in public expenditures, including in education. I do not believe that neoliberal policies will be changed very quickly and that all the money now spent on wars and on making the rich richer will suddenly change. It is important though to constantly remember why education around the world is in the state that it is in, and not to pretend that there is some solution to this within the educational arena alone (Berliner 2005). While we work hard to provide a high-quality education to everybody's children, we also need to keep pushing for a redistribution of the resources of this planet to more of its inhabitants and to the things that really matter. In the meantime, it is important not to lose sight of the fact that teachers are the most important factor in determining the level of educational quality whatever the resources at hand.

Currently educators and politicians often debate each other in dichotomies such as understanding teachers as technicians or teachers as reflective professionals or having teacher-centered instruction or learner-centered instruction as if one has to choose. We need to reframe the debates in ways that resolve the tensions inherent in positioning teachers in educational reform, addressing the legitimate role of both states and teachers. Neither an idealized form of teachers as reflective practitioners or technicians will ever be achievable; nor should they be. Teachers need to play an active role in shaping educational reforms and have the opportunity to adapt them to the varied circumstances in which they work, but they also need guidance and support in doing so.

With regard to the issue of learner-centered and teacher-centered instruction, it is likely that a hybrid model of teacher-centered and learner-centered methods such as the one emerging among BETD graduates in Namibia will be the most productive course to pursue. The history of educational reform in the United States (e.g., Cuban 1993) has shown that at best, efforts to create more learner-centered instruction in schools has resulted in the type of hybrid model of teaching that we and O'Sullivan have found in Namibia.

In the end, we should not settle for anything less for everyone's children, than we would want for our own children. "Good enough" teachers and "good enough" assessments of these teachers and their pupils should be "not good enough" for anyone's children.

Finally, there is an important role for teacher unions to play in this struggle for public education. Importantly, both the National Education Association in the United States and the Namibia National Teachers Union have challenged the narrow technocratic reasoning and deprofessionalization of teaching (e.g., Maletsky 2007), a path that the American Federation of Teachers has chosen not to pursue, for reasons Lois Weiner explores in her essay. Not only should teacher unions continue to combat the punitive attacks on teaching as a profession such as we have seen in the United States and Namibia, they should also join with teacher educators in the tertiary sector and defend the right of teachers to receive the kind of preparation for teaching that is necessary for teachers to enact their professional roles. The future of college and university-based teacher education and public schooling are closely linked, and it is time that teacher educators and teachers and their unions joined together to create the conditions that will provide a high-quality education to everyone's children.

References

Allington, R. 2002. *Big brother and the national reading curriculum: How ideology trumped evidence*. Portsmouth, NH: Heinemann.

Angula, N., and S. Grant-Lewis. 1997 Promoting democratic processes in educational decision making: Reflections on Namibia's first five years. *International Journal of Educational Development* 17:222–49.

Avalos, B. 2000. Policies for teacher education in developing countries. *International Journal of Educational Research* 33:457–74.

Baines, L. 2006. Deconstructing teacher certification. *Phi Delta Kappan* 88 (4): 326–29.

Barber, B. 1997. Public schooling: Education for democracy. In *The public purpose of education and schooling*, J. Goodlad T. McMannon, 21–32. San Francisco: Jossey Bass.

Berliner, D. 2005 Our impoverished view of educational reform. *Teachers College Record*. http://www.tcrecord.org/ (accessed August 19, 2005).

Burke, A. 1996 Professionalism: Its relevance for teachers and teacher educators in developing countries. *Prospects* 26 (3): 531–42.

Carnoy, M. 2000. Globalization and educational reform. In *Globalization: Integration and contestation across cultures*, ed. N. Stromquist and K. Monkman, 43–62. Lanham, MD: Rowman & Littlefield.

Craig, H., J. du Plessis, and R. Kraft. 1998. *Teacher development: making an impact*. Washington, DC: Academy for Educational Development.

Cuban, L. 1993. *How teachers taught: Constancy and change in American classrooms*. 2nd ed. New York: Teachers College Press.

Dillon, S. 2006. Schools cut back subjects to push reading and math. *New York Times*, March 28. http://nytimes.com/ (accessed August 28, 2006).

Goodlad, J., and M. F. Klein. 1970. *Behind the classroom door*. Worthingon, OH: Jones.

Grissmer, D. 1999. Class size effects: Assessing the evidence, its policy implications, and future research agenda. *Educational Evaluation and Policy Analysis* 21 (2): 231–48.

Guthrie, G. 1990. In defence of formalistic teaching. In *Teachers and teaching in the developing world*, ed. V. D. Rust and P. Dalin, 219–34. New York: Garland.

Haycock, K. 2005. Choosing to matter more. *Journal of Teacher Education* 56 (3): 256–65.

Ingersoll, R. 2003. *Who controls teachers' work: Power and accountability in America's schools*. Cambridge, MA: Harvard University Press.

Jansen, J. 1999. Does class size matter? The contribution of policy to the education crises of developing countries. In *International trends in teacher education: Policy, politics, and practice*, ed. M. Samuel, J. Perumal, R. Dhunpath, J. Jansen, and K. Lewin, 147–51. Durban, South Africa: University of Durban-Westville, Faculty of Education.

Lipman, P. 1998. *Race, class, and power in school restructuring*. Albany, NY: SUNY.

Maletsky, D. 2007. Teachers reject ETSIP plan. *The Namibian*, August 3, 2007. http://allafrica.com/ (accessed August 15, 2007).

Marope, M. T. (February,2005). *Namibian human capital and knowledge development for economic growth with equity*. Africa Region Human Development Working Paper Series, No.84. Washington, DC: World Bank.

McNeil, L. 2000. *Contradictions of school reform: Educational costs of standardized testing*. New York: Routledge.

Meier, D., A. Kohn, A., L. Darling-Hammond, T. Sizer, and G. Wood, eds. 2004. *Many children left behind: How the No Child Left Behind act is damaging our children and our schools*. Boston: Beacon Press.

National Priorities Project. 2007. A vote for more war. http://www.nationalpriorities.org/Publications/A-Vote-for-More-War-States-and-Congressional-Dist.html (accessed August 23, 2007).

O'Sullivan, M. 2004. The reconceptualization of learner-centered approaches: A Namibian case study. *International Journal of Educational Development* 24 (6): 585–602.

Peske, H. G., and K. Haycock. 2006. *Teaching inequality: How poor and minority students are shortchanged on teacher quality*. Washington, DC: Education Trust.

Pomuti, H., D. K. LeCzel, P. Swarts, and M. van Graan. 2006. *Practicing critical reflection in teacher education in Namibia*. Paris: Association for the Development of Education in Africa.

Reimers, F. 1994. Education and structural adjustment in Latin America and sub-Saharan Africa. *International Journal of Educational Development* 14:119–29.

Samoff, J. 1999. Institutionalizing international influence: The context for educational freeform in Africa. In *International trends in teacher education: Policy, politics and practice*, ed. M. Samuel, J. Perumal, R. Dhunpath, J. Jansen, and K. Lewin, 5–35. Durban, South Africa: University of Durban-Westville, Faculty of Education.

Samuel, M. (July 2005) *Accountability to whom? Or what? Teacher identity and the force field model of teacher development*. Paper presented at the 50th World Assembly of the International Council on Education for Teaching, Pretoria, South Africa.

Schwille, J. & Dembele, M. 2007. *Global perspectives on teacher learning: Improving policy and practice*. Paris: International Institute for Educational Planning, UNESCO.

Swarts, P. 2001. Teacher education reform: Toward reflective practice. In *Democratic teacher education reform in Africa: The case of Namibia*, ed. K. Zeichner and L. Dahlstrom, 29–46. Boulder, CO: Westview.

Tabulawa, R. 2003. International aid agencies, learner-centred pedagogy and political democratization: A critique. *Comparative Education* 39 (1): 7–26.

UNESCO 1998. *World education report*. Paris: UNESCO.

Valenzuela, A., ed. 2005. *Leaving children behind: How Texas style accountability fails Latino youth*. Albany, NY: SUNY Press.

Villegas-Reimers, E. 2003. *Teacher professional development: An international review of the literature*. Paris: UNESCO International Institute for Educational Planning.

Villegas-Reimers, E., and F. Reimers. 1996. Where are 60 million teachers? *Prospects* 26 (3): 469–92.

Walsh, K. 2004. A candidate-centered model for teacher preparation and licensure. In *A qualified teacher in every classroom*, ed. F. Hess, A. Rotherham, and K. Walsh, 223–54. Cambridge: Harvard Education Press.

World Bank. 2007. *Program document for a proposed first policy loan in the amount of US $ 7.5 million to the Republic of Namibia for a first education and training sector improvement program*. Report No. 38571. Washington, DC: World Bank.

Zeichner, K. 1999. Beyond an economic discourse on the quality of teacher education programmes. In *International trends in teacher education: Policy, politics, and practice*, ed. M. Samuel, J. Perumal, R. Dhunpath, J. Jansen, and K. Lewin, 68–71. Durban, South Africa: University of Durban-Westville, Faculty of Education.

———. 2003. The adequacies and inadequacies of three current strategies to recruit, prepare, and retain the best teachers for all students. *Teachers College Record, 105*(3): 490–515.

———. 2005 *Teacher education reform and educational reform in post-independence Namibia*. Final Report to the Spencer Foundation, Chicago.

Zeichner, K., and Luecke, J. 2004. *Teacher education as the basis for national educational reform: A case study of the reform of teaching and teacher education in post-independence Namibia.* Paper presented at the annual meeting of the American Educational Research Association, San Diego, CA, April 2004.

Note

1. This chapter is based on a keynote address presented at the 50th World Assembly of the International Council on Education for Teaching, Pretoria, South Africa, July 2005. Parts of this work were supported by the Spencer Foundation. This chapter does not express the position of the Foundation.

CHAPTER 14

Universalization of Elementary Education in India
A Dream Deferred Is a Dream Denied

Basanti Chakraborty

In 1950, long before goals were set by the international community, the constitution of India in Article 45 guaranteed free, universal, compulsory education for all children to the age of fourteen, with female education as a key element. However, despite numerous promises and five-year plans from governments up to and including the present, which made an unequivocal commitment in 2004 to provide at least 6 percent of Gross Domestic Product to education, the guarantee remains for many children a distant dream.

Some basic statistics show the extent of the need for a huge injection of resources into education. A recent survey found out that on an average, forty-one students share a classroom, nearly 10 percent of the country's elementary schools have only one classroom, and nearly 42,000 government schools have no building at all.

The level of literacy and school life expectancy for young children in India, and particularly for girls between the ages of six and fourteen, presents a bleak picture. According to a 2002 UNICEF report, India has over 19.5 million out-of-school girls, the largest number in the world. And among the groups that face the most discrimination, literacy rates for women were as low as 7 percent.

As a recent report demonstrates, motivation of parents in India toward educating their children is high, with 89 percent answering yes when asked if it is important for a female child to be educated. Despite parental desire, conditions sabotage their goal. India lacks school buildings and has an overburdened curriculum, overcrowded classrooms, scanty teaching aids, inadequate teacher training, and questionable accountability. All of these factors coalesce to deny children the opportunity for a basic education, let alone one that promotes the joy of learning.

India has received external funding from the World Bank and other international agencies for a program called Education for All or Sarva Shiksha Abhiyan (SSA). SSA aims to achieve the goal of Education for All by 2010 and 75 percent literacy by 2007. In this chapter I want to concentrate on the experience of SSA on the ground, which I have observed in many visits to India. SSA too often means frustration, despair, and mistrust about the extent and quality of its implementation.

Currently 91 percent of India's schools are in rural areas, and teachers' status is low. As one young woman primary teacher working in a rural area told me,

> Our position is the lowest in the education hierarchy. We are expected to bear the burden of education of rural children in isolation and often with no or very little support. We are expected to perform other functions such as census taking, etc., which take time away from our teaching responsibility. Most often male trainers train us. We do not feel comfortable with this—it is scary simply because in traditional settings we are not treated as equal with men. We are restricted from receiving training in far-off places because it is unsafe to stay overnight. Besides, it is hard to adjust the time and attend the training due to lack of fast transportation in rural areas. As a result we are barred from the professional development we need to receive. Even the bare provisions of water and toilet facilities are nonexistent in many places.

I also spoke to a woman teacher who lives in an urban area with her aging parents and her unemployed siblings and commutes every day to teach in the countryside, a pattern repeated all over India. She described the ordeal of her complicated commute to school: She rides on a rickshaw to the river, waits for a while for a boat to cross, and then walks about two miles on the village road to reach the school. Then her work begins:

> As I reach school every day, I cannot start to teach immediately. I have to take care of data collection, reports for the District office, arrange for midday meals and test the children on paper-pencil unit tests. I get very little time to teach my students. Besides, I do not have a toilet in school, and the villagers have no toilets of their own, either. So as soon as I finish my routine work, I hasten to come back to the riverside to catch the boat before it is too late, and then get the rickshaw to get back home and then take care of my aging parents. It is hard, but I wish we had better facilities, better accommodation for teachers,

and I could stay overnight and visit my parents over the Sundays and holidays! The school is held in a rented house. I have a blackboard, which the village leader keeps under lock and key at night. There are two other male teachers in the school. Both the men come to school [only] on days when the inspectors come to visit the school.

Recently I made a visit to a primary school in the heart of a city, with residential facilities for scheduled-caste boys from neighboring rural areas. The person in charge belongs to the scheduled caste also. The school was functioning in an incomplete cement building with open windows and doors, unfinished classrooms, and no water or electricity supply. Children carried their drinking water in buckets from the city supply. They sat on the uneven floor on floor mats, without complaint. Each classroom had a blackboard, but the rooms were dark except when sunlight peeped through the window. The teachers employed were mostly women. None of the teachers had any college education, and they had not received any training. The head of the school received some financial assistance from the SSA and philanthropic organizations and individuals. He moved from person to person, leader to leader in arranging money to buy food, clothing, and study materials for the children. Despite these hardships, the school benefits from being run by one person and is fairly secure and stable. The situation in government-run schools is very different, with management carried out at different levels, resulting in organized chaos and leading to frustration among teachers and parents and little joy of learning among the children.

One SSA supervisor I interviewed pointed, with pessimism, to serious flaws in the overall management and implementation of the SSA. He was originally a secondary school teacher, but after the inception of SSA was appointed as the supervisor of primary schools in rural areas. He tried to work with the system for a while but resigned from his job in frustration, citing lack of accountability on the part of higher administration and corruption among the top officials and politicians. He was very skeptical about the possibility of the SSA achieving the goals of education for all that it had set for itself. As he put it, "People like us who actually worked to implement the SSA at the grassroots level doubt how this plan can become successful without mobilizing necessary physical, financial, and human resources." He described the conditions I have noted above, adding the problem of teacher salaries. They receive less than $50 per month, and given the demoralizing conditions under which they work, it is not surprising that absenteeism is not considered a big crime. Teachers can appoint a substitute teacher to work for them on a contract basis. This contract teacher could be a high school dropout from the village. As a result, someone takes charge of young primary children who has no idea about education. He also observed that teacher training is sporadic and unorganized. Further, SSA inspectors are sometimes not experienced in primary education, so even when they do make recommendations to improve teaching, the ideas are not right.

A major obstacle to full education is that teachers who want to teach the children cannot do so because other chores take precedence. Primary teachers in India give surveys, collect census data, keep enrollment data, and in addition compile statistics on student enrollment, retention, and learning. As a result the primary school teachers have very little time to teach. Top level leaders of SSA consider data collection, not teaching or even building maintenance, to be the priority.

Corruption is a significant problem under SSA. Schools provide midday meals to primary school children, and in rural areas the teacher's entire time is spent in making arrangement to receive lentils, rice, and other food and finally to get someone to cook it. Private vendors supply the ingredients for midday meals as well as school uniforms, and because of top-level corruption, nobody objects to the low quality food or the substandard cloth used in preparing their clothing. Yet under these extremely poor conditions it seems as though the teacher's most important task is to administer a strict schedule of unit tests on six different occasions per year.

The views of this former SSA supervisor were shared by a rural activist who had dedicated his service to the poor people in his constituency over four decades. This gentleman works with poor villagers on a daily basis and helps high school students to get help to buy books and other materials to stay in school. He explained how poverty interferes with parental wishes for their children to be educated. Though India's poverty affects children of all religious backgrounds, the scheduled caste, scheduled tribes, and poorer Muslims are likely to be denied the educational opportunities of the SSA. Students in secondary schools drop out prior to finishing school because of poverty. Buying textbooks and clothes for high school students is extremely difficult for poor parents. Gender too is a factor. As he explained, "Poor parents cannot afford to send their girls to school for several reasons. Once the girls are mature, they want to give them in marriage. Educating their girls means paying a higher dowry for a more educated groom. Traditional parents do not want to send their girls to high schools in distant areas where there are no female teachers."

He pointed out that women from economically deprived classes and lower castes were mostly illiterate and untrained and had low social and economic status, and he argued that programs for women's empowerment were desperately needed to educate and train them to become economically self-sufficient.

In areas where women are empowered, he said, they take an active role in sending their children, especially girls, to schools, demand funding, take part in the midday meal program, and actively get involved in community governance toward the universalization of elementary education.

But above all this man expressed despair and frustration over the position of teachers under SSA, adding that the status of a country is known by the way it treats its teachers. "Unfortunately, in modern India, since the inception of the World Bank SSA program, the status of teachers has fallen to the lowest level." He cited low teacher salaries as the biggest bottleneck in executing the education plan. With such low morale, it is hard for teachers to exert their ethical responsibility as leaders of children in implementing the curriculum. "For the poor people of India, basic education is their only hope. Due to the lack of coherent results in receiving education in rural areas, people are experiencing a deep sense of desperation with respect to the capability of the state and nation to meet their educational needs."

As I listened to the stories, frustrations, and feelings of hopelessness of many people responsible for education in the country, I felt that the people of India, especially the poor, have been denied their fundamental right to educate their children. Education and the joy of learning is the right of every single child, and this is not secured by ticking boxes and data collection but by a properly trained and respected teaching force. The goals of Sarva Shiksha Abhiyan or Education for All will continue to remain a mirage for the rural poor in India until the importance and status of teachers is reflected in their salaries and conditions of service, until huge amounts of funding are put in to build the infrastructure of schools particularly in rural areas, and until the program is run by educationalists and democratically accountable governments—and not by those who see the education of the poor simply as a way of tying India's millions into the interests of supranational trade and banking institutions.

CHAPTER 15

Educational Restructuring, Democratic Education, and Teachers

Álvaro Moreira Hypolito

In this chapter I explain why the conservative restoration, which includes neoconservative and neoliberal perspectives, has attempted with relative success to impose its ideology in trying to create a new "common sense" (Apple 1996). Further, I will try to show how progressive movements might resist and fight for alternative practices in education. In my essay I use Brazil's experience, the use of political space created through a popular resistance to neoliberalism, to create public schools that involve parents and community, that rely on an emancipatory pedagogy, as an example of an historically "eye-opening" moment from which much can be learned.

Globalization is a combined and uneven process that can present different faces and characteristics according to the different situations where it is developed (Ball 1994). Contradictions between global and local cultures, dominant and dominated practices, and oppression and resistance tend to appear as a result of social and cultural dynamics. While we are living in times of capitalist globalization, as the essay by Susan Robertson (in Chapter 1) explains, we are also living the crisis of our time and its contradictions, despite capitalist vainglory. In the attack on progressive education and critical pedagogy, neoliberalism allies itself with the social conservatives who aim to impose their idea of morality and culture on the rest of the world. In my essay I focus on this "conservative alliance" between neoconservatism and neoliberalism, and the resistance to it, in educational restructuring. My focus assumes that one cannot separate the economic and political aspects of the neoliberal project in education from the cultural component of the "conservative restoration."

It is crucial to understand that there is also a strong progressive discourse, a counterdiscourse, which has been influential in teacher education and is present in schools resisting the dominant assumptions and ideas. As Brazil's experience exemplifies, educational restructuring on a local level reflects struggles between different actors with different sorts and amounts of power, as well as "discourses," or assumptions, the ways we think and discuss ideas. Local struggles and discourses are mediated by global realities and configurations.

Unlike conservative political forces that have been able to exploit control of governmental mechanisms and media, progressive movements in Brazil's cities and rural areas have exercised social and political power through a range of tactics. Actions include strikes, demonstrations, marches, occupations, and so on. In Brazil, teachers and scholars have been participating in the organization of teacher unions and of social movements, such as the Landless Movement, workers' unions, movements in defense of public schools and ethics in politics, and so on. In addition, critical pedagogy, seen in the work of the Brazilian educator Paulo Freire, has profoundly influenced Brazilian educational thought, as well as schooling internationally.

Similar struggles are taking place throughout the world, illustrating ways to construct an emancipatory perspective that contrasts with conservative ideology and policies. For instance, alternative publications, such as the progressive U.S. magazine *Rethinking Schools* and the book *Democratic Schools*, have been valuable instruments for teachers and parents who want to understand political aspects of education, fight for social justice, and create living examples of emancipatory education. In contrast to neoliberalism's assumption that "there is no alternative" to its ideology and global project, these publications show that contradictory notions, including the defense of public schools, the maintenance of radical conceptions of democracy, and the understanding that knowledge is socially constructed—that is, that people create knowledge—are ideas that are as natural as everyday life.

Globally, the most sustained challenge to neoliberalism and its educational policies has emerged in Latin America. In Argentina, teachers mounted a national strike for several months in defense of public schools, their funding, better labor conditions and wages for teachers, and so on. Teachers in Brazil have also fought in favor of public education and against conservative and neoliberal policies, and the teacher union has organized several battles for better salaries, a curriculum that makes students critical thinkers, and democracy. Teachers have resisted inside and outside school. Sometimes they have fought silently by not doing what official educational policies dictate; sometimes they have collectively resisted governmental offensives. While some might point to all that has been lost, it is equally the case that in the last twenty years, teachers have stood in opposition to official policies and, in many ways, deserve the credit for preventing the consolidation of neoliberal educational policies. When teachers, like any other group of workers, are fighting collectively, such as the Mexican teachers shown in the movie *Granito de Arena*, they can see things more clearly, and these special moments of struggle thus open eyes to new possibilities not otherwise envisioned.

Brazil's Experience: Popular Movements, Public Education, and Critical Pedagogy

Since the 1960s and 1970s, even under dictatorial regimes, interesting critical educational experiences have taken place in Brazil. Most of them were organized by political leftist organizations, non-government organizations (NGOs), progressive religious groups connected to liberation theology, and unions and were linked with social movements. In the 1980s, a new political scenario emerged with leftist and progressive parties ascending to important political positions. Many popular administrations have been installed in important municipalities led by an alliance called the Popular Front, which has governed major cities and several states. This political alliance included several progressive and leftist parties but has been led by the Workers' Party (PT). Recently, the political range of this alliance has been widened, and in many places the Popular Front has included some liberal and rightist forces (some of them part of conservative religious groups). In locales in which the Popular Front has won elections, progressive governments have been given the responsibility to give practical answers to public policies (public health, education, etc.), maintaining a focus on democracy and social justice under a hostile economic and political system.

Education is an important field in the battles for hegemony. Hence, progressive educational policies were emphasized and largely implemented by the Popular Front. For instance, in Belo Horizonte, the capital of Minas Gerais, a southwestern Brazilian state, the Popular Front has been implementing the Escola Plural ("plural school"), a program based on two general principles: education rights and construction of an inclusive school.

Education rights is the guarantee that all children have the right to go to school. This right is crucial when millions of children in Brazil have never gone to school. At the same time, it is a way to contest and minimize the exclusion process imposed by globalization. Although the dominant discourse of the World Bank and neoliberalism emphasizes that globalization is a process of inclusion, such as providing poor children who have been denied education access to schools, in reality, globalization imposed by transnational financial institutions can be seen as richocheting between inclusion and exclusion. Education rights are effectively realized only when they are translated into a pedagogical program. The second principle of the Escola Plural, developing an inclusive school, is absolutely related to the first, the right of every child to an education.

The Escola Plural's principles are articulated through curricular practices that understand "plural" as including many different factors: the construction of a school that considers education as a totality; schooling as a cultural experience and as a result of collective effort and production; schooling as a process in which the actors can redefine material conditions, both the funding and the physical circumstances of the classroom; schooling as an experience without interruptions; schooling as an adequate socialization process according to the different ages of the students; and schooling as a process that allows the construction of new identities of their actors—both students and professionals.

The practical repercussions of Escola Plural are intensive and radical. They impose new conceptions of knowledge, evaluation, and teacher education. This educational experience consists of changing public school education through a collective process including community, teachers, and students, based on practices that come from traditions of critical pedagogy. A central idea is the breaking of the correspondence between age and school level, rejection of the assumption that all people of a certain age "belong" in a certain grade or school level. In contrast, the Escola Plural reduces the eight years of elementary school to four or five cycles. Each cycle consists of two or three years, but the student can move to the next cycle at any time of the year, which allows students of different ages to be in the same classroom. Simultaneously, teachers and communities discuss how to include cultural aspects in curricular practices, especially in regard to groups socially and ethnically excluded, and how to exclude negative cultural aspects identified with the dominant culture.

The Escola Plural Contribution: Developing a Counterhegemonic Discourse in Education

A brief explanation of the role of ideology in politics and schooling will help clarify the Escola Plural's contribution to the struggle for emancipatory schooling. Given a specific historical moment, there is a dominant discourse that results from the battles and struggles between ruling and dominant groups or classes for control over cultural, economic, and political interests. The consolidation and solidification of the hegemony of the ruling groups are related to the capacity of the oppressed groups and their progressive movements to articulate and empower their counterhegemonic discourse(s). Neither ruling elites nor their progressive opponents are monolithic entities, and within both internal contradictions exist, in particular related to gender, religion, race, and social class.

For instance, working-class members often belong to conservative religious movements, which are identified politically with the conservative alliance, but at the same time the workers in these movements experience a different social and economic reality from the upper-middle class.

Social movements can use many of these contradictions to contest antidemocratic educational restructuring throughout the world imposed by neoliberal governments. While the conservative alliance often identifies its global interests with local realities, in reality, it imposes Western and capitalistic cultures. Despite the fact that there are peculiar and local characteristics in every educational system, conservative reform in education demands adherence to international and global standards (Hypolito 2005).

Still, it is important to acknowledge that as the conservative alliance forces its educational standards on diverse communities throughout the world, simultaneously, the dominant culture is not immune to the influence of the local cultures, an issued explored with great insight in the work of scholars in postcolonial literature. Local cultures can help fracture the cultural and political hegemony of the conservative bloc, and education plays an important role in these battles. On the one hand, the conservative alliance has been organizing its educational and political agenda to restore social values that have been weakened by progressive social movements, in particular those struggling for women's political, social, and sexual equality. On the other hand, progressive social movements have been experiencing important challenges to their counterhegemonic efforts to have schooling reflect an emancipatory perspective of society. I think that critically examining how deeply much neoliberal educational reform has been grounded in conservative approaches may be helpful to make us more politically active. In particular, progressive teachers, attacked by conservative ideology, have an interest in fighting against all forms of domination and exclusion but may need illustrations of concrete experiences to empower them toward a critical perspective.

A Counterhegemonic Experiment: The Escola Plural Program

Beginning in 1989, the Popular Front—a coalition of leftist parties including the PT (Workers' Party) and the PSB (Brazilian Socialist Party)—started winning elections at the municipal level. This crucial shift of perspective resulted in educational and political changes that have been the focus of much study. By the early 1990s, the Popular Front was governing important Brazilian cities, such as Porto Alegre, São Paulo, and Belo Horizonte among others. After the World Social Forum and its successful experiment with participatory budgeting, Porto Alegre became famous as one of the most important new political experiences throughout the world in terms of a process of participatory democracy. In terms of education, many researchers have studied the experience of Porto Alegre. For example, Guareschi (1998) has discussed the cultural life of youth in Porto Alegre's *favelas* (slums), understanding how youths have constructed their identities in the context of the Citizen School in Porto Alegre. Gandin (2002) has dedicated his studies to analyzing the Citizen School as an educational and political program, showing us how it has been possible to creatively build alternative programs that challenge neoliberal reforms.

The Escola Plural Program was conceived and implemented in a similar context. In 1992, the Popular Front had begun its first term in the city of Belo Horizonte. It was a time of great hope for social movements and for the working classes. In particular, it was a time of hope for teachers and educators who had been engaged in the difficult struggle to create better conditions for public education in Minas Gerais.[1] The teachers' movement and the teacher union have played crucial roles in the historical battles for democratic management, increased funding for public education, community participation, inclusive education, better working conditions, and so forth. The Escola Plural Program comes out of this political and cultural heritage.

One of the ways in which hegemonic alliances have been built is through denouncing various crises in public education, such perennial problems as grade repetition and high dropout rates, mainly among working class and other excluded groups of students. Although official educational policies have followed conservative orientations and have been under ruling class control for many decades, sometimes it has been difficult for progressive social movements and teachers to acknowledge these problems with public schools. For example, schools often teach in ways that are completely disconnected from students' cultural contexts and their aspirations.

The Escola Plural Program has been an attempt to openly confront the problems of public schools. There is a consensus about the need for funding public education even though controversy surrounds the issue of how that educational funding should be invested. The Escola Plural has not, in fact, focused on funding. Rather it has focused on pedagogical and political concerns. It is not a million-dollar program, such as the Pro-Quality Program, a managerial reform designed by the World Bank for "helping and improving" the educational system in the State of Minas Gerais, Brazil (Hypolito 2005), but a much more humble program in terms of financial investments. It is both simple and creative, although this does not mean it is simplistic (Secretaria Municipal de Educação 1998, 2002). I should note that although the Escola Plural program is still active, some people involved in its creation see distortions in its operations at present.

The Escola Plural Program represents a decision to form a collaboration between those who are involved in the educational process and those for whom it is intended. Key aspects are support for teachers to have more autonomy and time for planning and studying, democratic school governance, and elections for principals.

The Escola Plural as a program for municipal education started in 1995, when it was implemented in 173 municipal public schools in Belo Horizonte. It was implemented over the period of 1995–1997, involving over 146,600 students in "fundamental," that is, elementary and middle, school, and 9,700 workers in education, including teachers, principals, supervisor, and secretaries.

The Escola Plural has tried to radically transform schoolwork organization by means of reorganizing time and changing the labor process for teachers and students. In terms of curriculum, the Escola Plural Program has tried to break with traditional modes of teaching, rejecting concepts of teaching focused on the teacher as transmitter of knowledge. The Escola Plural Program has also been an attempt to reduce grade repetition and failures, eliminating mechanisms through which exclusion has been accomplished, and replacing them with an evaluation system that involves a discussion about what to evaluate, who evaluates whom, and how. The Escola Plural has been concerned with transforming the relationship between those who have been traditionally the subject of reforms (students and teachers) and the knowledge they are supposed to learn. This change is attempted by understanding school knowledge as related to local and global contexts, by overcoming a fragmented organization of disciplines, and by incorporating critical traditions and local culture.

In its process of implementation, the Escola Plural Program has aimed to redefine schooling through these means:

- A collective intervention has been made to radically reduce school repetition, failure, and dropping out through trying to build a more democratic and egalitarian public school.
- Human education has been understood as a totality. The Escola Plural Program conceives of schools as a plurality of social and cultural spaces

and times in which students and teachers socialize, teach, learn, and construct their identities.

- School has been made a time and location of cultural experience that connects its actions with communitarian cultural production. Culture is understood as life and not as a curricular discipline.
- School has been made a collective production with discussion groups, collective production of instructional materials, textbooks, collective governance, discussion of educational ends and politico-pedagogical proposals, and other collaborations.
- Schooling's material conditions, such as facilities and work conditions, and some virtual aspects, such as organization of time, spaces, labor process, schedules, etc., have been understood to limit innovations. Spaces, time, rituals, logics, disciplines, grades, and contents all have to be radically questioned.
- There has been an insistence on age-appropriate learning experiences and respect for childhood and children in their self-images, identities, rhythms, languages, representations, etc.
- Adequate socialization for each age-cycle or group of instruction has been unimpeded by differences of race, gender, class, learning rhythm, etc. (I discuss the construction of race in Brazil elsewhere [Hypolito 2004].)
- A new professional identity has been built for new professionals who are much more plural and who participate in processes of democratic school governance, workshops, interdisciplinary projects, participatory planning, research, collective and collaborative production, and a rich process of in-service education. Paid time for studying and researching is another expression of this new professional consciousness.

Within the space I have in this essay I cannot explore fully all of these elements of the Escola Plural Program, but I will explain one core aspect to indicate how far reaching and how transformative this project is: the reorganization of time in schools. This refers to a concern about how time is used in schools. What should be the logic of temporal organization? Organizing and managing time in schools is one of the most conflictive aspects, as seen in the school calendar, class distribution, exams, holidays, recess, and the timetable. This is an area of tension because it deals with private and personal interests. Time in schools is work time, paid or unpaid time, time for relaxing or time for pressure and stress. Time involves teachers' and students' rights. Teachers' movements and their unions have been fighting to gain more control over how time is managed in schools.

According to the Escola Plural Program, in the dominant (and assumed) logic, time is organized according to "average rhythms" of learning and does not take into account cultural diversity or differences of gender, class, and race. It supposes "simultaneousness": a student has to master material simultaneously in all disciplines. It assumes specific times for specific abilities and skills: time for reading, time for writing, time for mathematics, etc. It assumes constraints of time. This logic assumes a separate time for literacy, mathematics, library, the arts, a time for each discipline; it separates time to manage, time to teach, time to learn, time to evaluate, etc. Yet, this can be seen as a logic of exclusion that has had perverse consequences for the working class and minority groups.

Encouraging experiments and learning from them, the Escola Plural Program articulated a proposal to reorganize time in Escola Plural. Accordingly, time is understood in a flexible way, and the structure of grades (first grade through eighth grade) has been radically redefined, shifting traditional grades to cycles of education, organized as cycles based upon age.

Grades are organized into three cycles of three years each, which the Escola Plural Program proponents (Secretaria Municipal de Educação or SMED) believe adequately accommodate students' sociocultural diversity, different levels of maturity, and different learning rhythms. Within each cycle, age must be considered in the organization of classes. This new organization of time assumes that students will continue studying with their age group peers without rupture, interruption, or repetition. It is possible that someone may not be well prepared to be promoted to next cycle, but such a decision will only be made after a qualitative, descriptive, and collective evaluation.

Figure 15.1 Organization of Escola Plural Basic Cycles

Cycles	Stage	Ages	Age Groups
First	Childhood	6, 7, and 8/9 years	6–7 years 7–8 years 8–9 years
Second	Preadolescence	9, 10, and 11/12 years	9–10 years 10–11 years 11–12 years
Third	Adolescence	12, 13, and 14/15 years	12–13 years 13–14 years 14–15 years

(After SMED, 2002)

Finally, the last aspect of the reorganization of time in schools refers to the reorganization of professionals' time. Reorganizing school time implies reorganizing teachers time, for example their life, family, other jobs, and qualifications. With this in mind the proposal defined some aspects to orient discussions, while intending that teachers themselves should define the reorganization of time collectively. The aspects to be considered included preserving all teachers' rights, such as vacations, contracts, time for studying and planning, and collective school governance; organizing time for nonclass activities, such as coordination, supervision, management; creating a collective of teachers for organizing and planning each cycle and area of knowledge.

Many emergent and emancipatory practices were taking place throughout the public municipal educational system in Belo Horizonte, and like the Escola Plural program, most were critically oriented, and many followed Freire's ideas. Freire emphasizes that teaching cannot be separated from cultural and political struggles: "Teachers should struggle with the dominant cultural values that are present both in the society and inside themselves in order to understand their cultural and political function. . . . And, once again, such transformative work would necessarily go beyond the classroom" (Apple, Gandin, and Hypolito 2001, 130).

Figure 15.2 visually represents how the Escola Plural understands curriculum and its relation with the disciplines. Drawing on Freire's notions, the curriculum emerges from themes, rather than being confined to what are often arbitrary limits of a subject or discipline.

The last core aspect underpinning the Escola Plural Program refers to assessment. Considering school evaluation as key for any innovative pedagogic proposal, the Escola Plural Program has elevated evaluation to a high level of concern. Assessment is discussed by teachers, schools, students, and parents. Most have in mind questions such as: What is to be evaluated? For what ends? Who will do this? When? And how should this be done? The

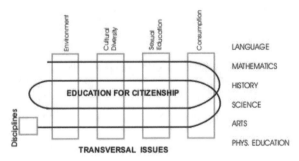

Figure 15.2

Escola Plural Program has attempted to answer these questions by working with teachers and schools and stimulating new forms of evaluation: portfolios, qualitative and descriptive assessments, and cooperative evaluation, among other forms. The key aspect is to create alternative modes of assessment in which all participants—teachers, students, and parents—take part in the process.

References

Abers, R. 1988. From clientelism to cooperation: Local government, participatory policy and civic organizing in Porto Alegre, Brazil. *Politics & Society* 26(4): 511–37.

———. 1996. From ideas to practice: The Partido dos Trabalhadores and participatory governance in Brazil. *Latin American Perspectives* 23 (4): 35–53.

Apple, M. W. 1996. *Cultural politics and education*. New York: Teachers College Press.

———. 2001 *Educating the "right" way: Markets, standards, God, and inequality*. New York: Routledge-Falmer.

Apple, M. W., L. A. Gandin, and A. M. Hypolito. 2001. Paulo Freire, 1921–97. In *Fifty modern thinkers on education—From Piaget to the present*. London: Routledge.

Avritzer, L. 1999. Democratization and changes in the pattern of association in Brazil. *Journal of Interamerican Studies and World Affairs* 42 (3): 59–76.

Baiocchi, G. 2001. Participation, activism and politics: The Porto Alegre experiment and deliberative democratic theory. *Politics & Society* 29 (1): 43–72.

Ball, S. 1994. *Education reform: A critical and post-structural approach*. Buckingham: Open University Press.

Bernstein, B. 1996. *Pedagogy, symbolic control and identity—Theory, research, critique*. London: Taylor & Francis.

Gandin, L. A., and M. W. Apple. 2002a. Challenging neoliberalism, building democracy: Creating the Citizen School in Porto Alegre, Brazil. *Journal of Education Policy* 17 (2): 259–79.

———. 2002b. Can education challenge neoliberalism? The Citizen School and the struggle for democracy in Porto Alegre, Brazil. *Social Justice* 29 (4): 26–40.

Guareschi, N. 1998. The favela and the school: Contradictions and resistance in student's construction of identities. PhD diss., University of Wisconsin–Madison.

Hypolito, A. M. 2004. *Global educational restructuring, school organizations, and teachers: The effects of conservative and counter-hegemonic educational policies on teachers' work in Brazil*. PhD diss., University of Wisconsin–Madison.

———. 2001 Multiracial reality, white data: The hidden relations of the racial democracy and education in Brazil. In *Global constructions of multicultural education: Theories and realities*, ed. C. Grant and J. Lei, 159–173. Mahwah, NJ: Lawrence Erlbaum.

Santos, B. S. 1998. Participatory budgeting in Porto Alegre: Toward a redistributive democracy. *Politics & Society* 26 (4): 461–510.

Secretaria Municipal de Educação. 1998. Escola Plural. http://www.pbh.gov.br/smed/escoplur/escplu00.htm (accessed May 12, 1998).

———. 2002 Escola Plural—Proposta Político-Pedagógica. Caderno 0, 2a. ed., Belo Horizonte, Outubro.

Note

1. When I was in the Education Graduate Program of Federal University of Minas Gerais I had the opportunity of knowing teacher union activists and radical scholars committed to political and educational reforms. Many of them were the leaders of Escola Plural. Many times in my professional life, as a teacher in public schools or as a professor, I have been engaged in the union as a member of the board of directors, as an activist, or as an ordinary member. That explains in part my knowledge of and my interest in investigating Escola Plural.

PART V

Neoliberalism, Inequality, and Teacher Unions

CHAPTER 16

Sodexho in the Chicago Public Schools

Kyle Westbrook

Most in my adopted hometown of Chicago will remember the summer of 1995 as the summer of the deadly heat wave that conservative estimates say cost 465 Chicagoans their lives. For eight intense days temperatures hovered between nighttime lows in the upper 80s and daytime highs up to 106 degrees Fahrenheit. I remember that summer as the time I was hired to teach at Lincoln Park High School. It was, as well, precisely the time the Chicago Public Schools (CPS) hired its first CEO, Paul Vallas. In fact, on the July morning that I went to the Byzantine world of the Chicago Board of Education headquarters on Pershing Road to be "staffed," the term that's used to describe how Chicago teachers are placed in their schools, news trucks and photographers were there to announce the mayor's appointment of Vallas, a corporate executive and his former chief of staff, to the top post in the CPS. What became apparent only in subsequent years was that the death toll from the heat wave wasn't the sole tragedy unfolding in Chicago that summer. Equally dangerous has been the way that Paul Vallas and his successor Arne Duncan have remade Chicago schools, part of the elite's attempt to transform Chicago into a global city.

To be sure, the neoliberal experiment in the Chicago Public Schools was under way before 1995; however, with Vallas at the helm, the movement gained considerable momentum. Business groups like the Commercial Club of Chicago had in Vallas a partner in reshaping the school system to meet its antiunion, market-based agenda. Even at a "good" school like Lincoln Park, by 1996, high-stakes testing in a range of subjects, from history to biology, dominated curriculum discussions and department meetings. Vallas ushered in a period of intensive high-stakes testing, a weakening of the Chicago Teachers Union, rapid privatization of school services, and an overall widening of the achievement gap between the system's highest and lowest achieving students.

One of the most basic propositions that has guided the neoliberal transformation of the Chicago Public Schools is an assumption that the Chicago Teachers Union has hurt the city's overwhelmingly minority, low income students through the union's narrow pursuit of higher salaries and better benefits for its members. I'm an admitted partisan when it comes to unions. When I was young, it was the International Brotherhood of Electrical Workers that won my mother her job back when everyone else seemed powerless to stand up to "Ma Bell" (the nickname given to the telephone company) on my mother's behalf. As I was sympathetic to unions, it's not surprising that as a young teacher I gravitated toward the union crowd at Lincoln Park.

Throughout my first year teaching a group of students that weren't welcome in anyone else's class, no administrator bothered to observe me. Aside from one colleague, the only people that did seem to give a damn about me as a new teacher and what type of learning was happening with my mostly African-American and Puerto-Rican students were the union people who regularly asked me how things were *really* going. Because we shared classrooms in Lincoln Park, the library workroom became the default teachers lounge for people interested in either working or talking to other teachers about teaching and not simply complaining about the students.

My first year was a dramatic one in CPS. The Vallas regime shook up the culture of CPS by generating headline after headline about waste at central office, corrupt principals, misguided local school councils, unqualified teachers, and, of course, a culture of failure in the schools. To a public fed up with systemic failure in the city's schools the headlines served two purposes: first, they convinced the public of the need for radical change in the system, and second, they cultivated the view of Vallas as a crusader for children. The implicit message that emanated from the board of education was that all of us that worked for the students had failed them and that what was needed was an outsider's view and more market-based measures.

Within months of taking office, Vallas created a new Office of Accountability to hold principals and teachers accountable for decades of failure. In a frenzy of accountability, CPS created two new standardized exams to gauge student performance. In spite of the criticism that both tests received from teachers and academics, CPS proceeded to use the tests to identify failing schools and in some cases "reconstitute," them, a top-down process to impose a new organization, student population, teaching staff, and curriculum, without input from parents, teachers, or students.

Another key component of the neoliberal agenda in Chicago has been the introduction of choices for parents who wish to send their children to schools other than the neighborhood school. The case of Englewood High School is fairly indicative of school "choice" in Chicago Public Schools. Englewood High, situated in the economically depressed, high-poverty neighborhood for which the school is named, was summarily closed due to its perennial low standardized test scores. Just weeks after the school closure was announced, against the protests of community members, the school was slated to reopen as Urban Prep Charter High School. Englewood faculty and staff were removed, though they were allowed to reapply for their positions in the new nonunion charter school.

Because the school in which I taught, Lincoln Park, has a prestigious International Baccalaureate (IB) program that attracts largely middle-class, high-achieving students to the school, the specter of reconstitution was never a serious threat. Though the IB program was relatively small in terms of the overall student body, the presence of that program in the school attracted other academically talented students to the school, creating a very diverse mix of "community" students, some who lived in the Cabrini Green housing projects located about a mile south of the school, as well as working-class and middle-class students. Students seemed to coexist peacefully amid relative segregation within the school. Yet faculty at Lincoln Park experienced daily assaults on our morale through the near-constant news stories and communications from the central office about school failure.

In my second year of teaching, I faced class sizes upwards of thirty-four, which violated our contract. Contrary to the advice of some in the school, I filed a grievance. I saw that the union was the only insurance I had to protect my ability to effectively educate my students, who tended to be from Cabrini and the surrounding area or other low-income communities around the city, and were in need of as much attention, particularly to their writing, as I could give them. Winning a class-size reduction through this union grievance made a significant difference in the amount of time I could devote to each student, time that was still pitifully inadequate.

Vallas and the school board began a massive program of privatizing school services within the first year of assuming control of the schools. The school reform act that ceded control of the CPS to the mayor also gave the board of education legal authority to terminate all nonteaching and support staff union contracts. Everyone from CPS electricians to cafeteria workers were replaced with private companies contracting with the board of education. By 1999, Sodexho Marriott had become the exclusive food service provider for the school system. In a city where school breakfast and lunch had been the most nutritious meals of the day for thousands of low-income students, the low-grade, fat-drenched, fried foods now being served by Sodexho were a cheaper but less healthy alternative to the relatively balanced menus served up by public school cafeteria workers pre-Vallas. To date there has not been any serious discussion about what has been lost with the outsourcing of many of these services, nor has there been a vigorous inspection of the quality of work and food sold to CPS. If a foreign power were to heap the breakfast and lunches that are routinely served up to students in the Chicago Public Schools on our suburban counterparts, it would likely be viewed as an act of war.

By 1999, having seen good teachers leave for the greener pastures of the suburbs or be chased out of the school because of their personal politics, and after witnessing the harassment of teachers by well-meaning but nonetheless pressured administrators, I decided that I should work to make life for teachers in the school a little better. In 1999 I ran for and was elected to represent Lincoln Park High School to the citywide Chicago Teachers Union House of Delegates. At my first meeting I spent two hours watching the autocratic union leadership belittle and silence its critics. It became all too clear that part of the reason that teachers were in such a sorry state was that our union leadership seemed far more interested in staying in power than working to slow Vallas and neoliberalism's steamrolling of the profession and the rights of the membership.

Since 2001 the union has experienced the election of a reform president, continuing shifts of power among contending caucuses and a struggle with the Board of Education about how to improve schools that are considered failing. While the union has managed to win some reforms that are important steps toward improving instruction in this city's schools and are slowing the process of school closures, it's my contention that they don't go far enough. The Chicago Teachers Union has failed to use its position as one of the largest teacher unions in the United States to build coalitions with other cities and develop a coherent, long-range strategy for combating the privatization of the public schools. One failing is especially important: The union has failed to think creatively about how to address the chronic problem of teacher vacancies, particularly in schools serving low-income families. Every year the Chicago Public Schools faces hundreds of vacant teaching positions usually staffed by substitute teachers. While the Board of Education has devoted millions of dollars to alternative teacher certification programs and charter schools, little attention has been devoted to maintaining a stable faculty presence in hard-to-staff schools. If both the union and the Board of Education are to approach the chronic teacher shortages in the CPS's most troubled schools, new thinking and programs are needed to recruit and retain seasoned teachers.

Still more dangerous changes are in the pipeline. In 2004, Arne Duncan, Vallas' successor, announced Renaissance 2010, a dramatic plan to create one hundred new schools by 2010, all outside of the confines of the collective bargaining agreement. This initiative has left the Chicago Teachers Union baffled, unable to do more than make periodic public statements in opposition. In fact, it has been community organizations and another public employee union, the Service Employees International Union, that have been most vocal in arguing for an end to the plan. In February of 2006, a local politician introduced a resolution in the city council calling for a moratorium on school closures and serious analysis of Renaissance 2010. Whether the resolution and the growing chorus of voices opposing school closures will lead to a shift in policy remains to be seen. However, the relative silence on the part of the Chicago Teachers Union on the chartering of public education in Chicago has left many within the leadership and rank and file bewildered.

While those of us fortunate enough to have been around when CPS employees ran the cafeterias bemoan the quality of the food served by Sodexho Marriott for breakfast and lunch every day in Chicago school cafeterias, the fact that a private, for-profit company is profiting off of school children in our schools seems to elude public debate and scrutiny. The lack of public voice in Chicago of one of the most obvious stakeholders in public education, its teachers, demonstrates the difficulty teachers confront in fighting the "good sense" of the neoliberal agenda in the Chicago Public Schools. A clear, unambiguous vision for change in the CPS that places the interests of the city's most vulnerable students at the center of the union agenda articulated by the Chicago Teachers Union is not just a prerequisite for system-wide improvement of the schools. It may be the last hope for the union and public education in the city to survive.

CHAPTER 17

Homophobia in St. Lucian Schools
A Perspective from a
Select Group of Teachers

Urban Dolor

Homophobia is defined by Hansman (2006) as "a prejudice similar to other manifestations of hate including sexism and racism." On the island of St Lucia these expressions are manifested through actions such as taunts, jokes, and physical abuse. On account of the negative reactions gay students face within the education system, it is reasonable to expect that they are likely to stay "in the closet" and to suffer all the attending negative repercussions associated with such a closed existence. The difficulty that such students face is likely to be exacerbated because the Roman Catholic Church exerts a domineering influence on the education system in St Lucia.

The church's dominance is confirmed by the fact that well over 75 percent of primary schools in St. Lucia are deemed to be Roman Catholic Schools. Among other things, this means that the Roman Catholic Church is responsible for appointing the manager for these schools, for approving the schools' principals, and for the religious instruction provided at these schools. What's more, until as recently as thirty-five years ago, the only secondary level education available to the overwhelming majority of St. Lucians was provided by the St. Joseph's Convent and St. Mary's College, both Roman Catholic schools. To date, these schools continue to be the premier secondary schools in the island.

The Roman Catholic Church's position against "the intrinsic evil of homosexual acts and the objective disorder of the homosexual inclination" has been documented and analyzed elsewhere (Ontario Consultants on Religious Tolerance, n.d.; http://www.religioustolerance.org/hommarbrcc.htm). Given that the church has been responsible for forging the religious (and thus moral) perspectives of most St. Lucians, particularly those who are more educated, it is easy to conclude that the overt display of nonheterosexual behaviors would not be tolerated in St. Lucian schools.

In an effort to obtain an indication of the extent to which homophobia exists in the classroom, a study was conducted in which forty-five St. Lucian teachers were asked to indicate whether they agreed or disagreed with twelve statements extracted from the article "Education and Queer Youth" by Glen Philip Hansman.

The selected teachers were all reading for a Masters in Education degree from the University of Sheffield. This program is a joint venture between the St. Lucia Teachers' Union and the university and represents one of the union's efforts to improve the quality of teachers in St. Lucia.

It is significant to note that less than 5 percent of teachers in St. Lucia have been educated at the Masters Level. This means that many of the persons who participated in the survey are likely to hold leadership positions within St. Lucia's education system within the next few years. We can conclude, therefore, that trends distilled as a result of the analysis of the response offered by these teachers can provide an indication of the policies that will guide the curriculum within the next few years. There is another compelling reason for using this sample of well-educated teachers. There is a generally accepted principle that more educated people are less likely to be prejudiced against others. This suggests that the homophobia observed in this group of teachers is likely to exist in even more virulent strains within the general population of teachers and within the education system.

The data indicate that participants support the proposition that gay and lesbian rights should be discussed in classrooms. In addition, almost 50 percent of the respondents held the position that issues around sexuality should be discussed in all areas of the curriculum. Unfortunately, the stark reality is that within St. Lucian classrooms there is no evidence that indicates that such issues are a part of the curriculum. This is perhaps understandable when you take into account that almost 50 percent of the sample agreed that teachers are resistant to strategies that seek to end hostility toward gay youth in the classroom. Even so, all persons who completed the questionnaire agreed that school policies should give all students, irrespective of sexual orientation, confidence that they will be protected from violence and discrimination while at school. Yet 20 percent of the respondents did not believe that teachers have an obligation to protect gay students from discrimination—this is cause for concern. When teachers are reluctant to accept changes that are meant to better the life of an important group of students within our schools, we must remain concerned about the likelihood that the suffering of these students will be addressed in any meaningful way.

A significant majority of the respondents indicate a clear intolerance for sexism and racism, although it is regrettable that even a small percent of persons did not share this view. However, it is heartening to note that three-quarters of the teachers did not always assume that boys should date only girls and girls should only date boys. It would be interesting to find out how teachers would react to real gay or lesbian relationships in their school communities.

The vast majority of respondents indicated that it is easy to identify gay students. This response does not take into account the sophisticated approaches that gay people utilize in order to avoid being "outed," and when it is coupled with the admission that educators do not deal well with topics related to sexuality, one can easily conclude that it is unlikely that teachers are in fact able to identify the majority of gay students.

The level of homophobia in the classroom is not likely to be any better on other Caribbean islands. After all, the religious and historical factors are similar. Indeed, evidence is that St. Lucians are more likely to be tolerant of sexual activities that may be deemed nonpuritanical on account of the influence of the French components of our history and the geographical proximity of Martinique.

The evidence from the data collected is that homophobia continues to exist in the education system in St. Lucia. This is hardly surprising given the prevalence of homophobia in St Lucian and Caribbean society as a whole. The evidence is unambiguous. For example, in St. Lucia within the last two years two young men were killed because they were gay. I have never heard anyone suggest that a person was killed because he was heterosexual, yet St. Lucians apparently find it easy to accept that a person could be killed for being homosexual. It was as if homosexuality was some scourge that had to be eliminated from the society. Other evidence of homophobia in Caribbean society abounds: the messages in the songs, the derogatory use of words that refer to gay men, the shame and the denial that families of gay men endure— I could go on and on.

In addition, my personal experience is that persons who speak out against homophobia must also deal with the kind of darts that are hurled at gay students. For example, I was the target of a number of jokes from many persons including the Minister of Education, who remarked that he heard I was an advocate for "them bullars" (a derogatory term for gay men) because I argued, via a television commentary, that homosexuality was a phenomenon that occurred naturally in human populations. This happened when I was president of the St Lucia Teachers' Union (SLTU). The *Star* newspaper carried the front page headline the next day "Union Prez says Homosexuality Natural."

Even government ministers are not insulated against the wrath of homophobes in this society. When the Minister of Health mooted the idea that homosexuality should be decriminalized as a step in the fight against HIV infection, he was forced to clarify that he was speaking on his own behalf and that his views did not necessarily represent the government's position. Against this background it is perhaps not surprising that the SLTU has never mounted a campaign to tackle the issue of homophobia in the classroom or in society. The general thinking within the union executive is that any effort to speak for the rights of homosexual people would not represent the views of the general membership. Under these circumstances the union has remained resolutely silent on the matter.

The Caribbean Union of Teachers (CUT) has not yet seen the need to advocate for the rights of gay people either. In fact, during my tenure as First Vice President of the organization I was consistently opposed whenever I took the position that the Union should have a clear position that homosexuals should be afforded the same respect and dignity as heterosexuals. Indeed, my bid to become president of the CUT was railroaded primarily on account of my pro-LGBT (lesbian, gay, bisexual and transgender) perspective. The fear was that the CUT would be portrayed as an organization that favors homosexuality.

We must accept that while educators are unable to completely eradicate homophobia in the classroom, it is our moral duty, as instructors in a democracy, to protect and give voice to our homosexual students and that teacher trade unions should be promoting this agenda. The survey I have described above does offer some grounds for hope that attitudes may be changing. I am also conscious that these matters have only become accepted as the legitimate concern of trade unions in many other parts of the world including the United States and the UK within the last couple of decades and that there too there was virulent opposition to the struggle for the rights of LGBT people. At the Fourth World Congress of Education International (EI), which I attended recently, there was a daylong symposium on these issues, and EI undoubtedly has a role to play in helping those unions such as my own to face up to their responsibilities in fighting for the rights of LGBT people, even against such a difficult culture in this respect as that of the Caribbean.

References

Hansman, G. P. n.d. Education and queer youth. Gay and Lesbian Educators of B.C. http://www.galebc.org/EDUCATIONANDQUEERYOUTH.pdf (accessed 7 August 7, 2007).

Ontario Consultants on Religious Tolerance. n.d. http://www.religioustolerance.org (accessed August 7, 2007).

CHAPTER 18

Work on Aboriginal Education in a Social Justice Union
Reflections from the Inside

Chris Stewart

With the support of the British Columbia Teachers' Federation (BCTF), the Task Force on First Nations Education 1998–99 developed and presented fifty-six recommendations for the BCTF Executive committee and the Annual General meeting 1999. Recommendation 1.4 asked that the BCTF hire a person of Aboriginal ancestry to the professional staff, with members of the BCTF of Aboriginal ancestry included in the shortlisting process, recommending who should be hired. The recommendations aimed to set the table for First Nations students, First Nations teachers, First Nations parent, and public education to sit together to create, facilitate, and nurture First Nations children and their families and community. For the first time all involved in public education appeared to acknowledge the devastating impacts colonization has had and continues to have on our children and communities. Union local leaders were to respond, strategize, and accommodate for change, that is, alter, the way they understand Aboriginal First Nations teachers and Aboriginal First Nations children.

In this essay I comment on experiences I have had in trying to move this agenda forward. I am a former classroom teacher and BCTF member, the second person to occupy this space in the union. In my work on this issue of social justice, I attempt to bring to light important issues that have long plagued the public education system and involvement of Aboriginal people as participants in public education. My role is to facilitate ongoing work on

First Nations concerns and mentor new activists into the union. I work with a nine-member advisory committee. When I started the position, I was met with a budget that was inadequate and a promise that if members needed money for projects, all I had to do was ask. One practical lesson I have learned is that needs have to be stated and pressed, that the existence of the need itself is no guarantee of the need's being satisfied.

The first year of its existence our committee began the process of investigating employment equity for Aboriginal teachers, and the struggle for recognition and funding continues. The issue arose within the BCTF in 1997, before formation of our committee, with a recommendation about "pull-out" for Aboriginal students in the province. This recommendation was very controversial, and everyone with an opinion about Aboriginal education and aboriginal learners wanted to say something. A BCTF member from one of the urban teacher locals asked that a task force on Aboriginal education be formed. A call for Aboriginal teachers to form the First Nations Task Force on Aboriginal Education was sent out. This was the first time that Aboriginal Education was put in the hands of First Nations and led to creation of our committee.

Challenges for Unions

There are a number of challenges I see in how a union committed to social justice, as the BCTF is, can hold simultaneously ideals of unionism and social justice in public education. Aboriginal teachers and our children are used as a starting point to have a dialogue on social justice, but the problem is not one that arises from Aboriginal issues. Rather, the problem is that of inequality and discrimination in the society and schools. If we are a union of substance, we need to put our truths on the table. This is not always the case, and within the BCTF teachers have not always seen eye-to-eye on social justice when it comes to the First Nations Task Force and their recommendations. Too often the work that is given to Aboriginal Educators remains separate from the perceptions of most members.

Questions we must answer: How can we move forward and balance the desires of the Aboriginal people and the desires of the membership? Who benefits and at what cost? Why is there such a discrepancy of or between minds? What criticisms can be articulated and which cannot? How can those who experience continuing discrimination maintain trust in a union's commitment to social justice and yet deal with what seem to be empty promises? I hear often "Just wait—once we have our needs addressed, we will address Aboriginal education." How do we do this, embracing all the issues?

The argument that first we must attend to the rights of the whole and then later to those who are excluded by discrimination from the rights enjoyed by the majority weakens the union. What if we were to focus on social issues as an issue "of the whole"? What would happen? We would be still paying dues but also supporting the idea among our membership that you are only as strong as the weakest link. This concept to me is about social justice. We do, however, have members who do not see this as a viable way to go and want the union to go in a direction that we see as only a focus on worker rights and benefits.

If we were to go down this path, we would have to forfeit the illusion that we care for the children we teach and ignore the fact that not every child in public education comes from places of privilege. While union members would all probably agree that we need to find and desire a better place for our children, how we get there is another question. Yet, how we interpret or live this question defines any union, including the BCTF.

To date in BC we have about 275 to 300 self-identified Aboriginal members in the BCTF. This number fluctuates, and we have experienced a decline over the years. The pressures are many for Aboriginal teachers. They are more often than not called upon to be the "cultural brokers" for all the aboriginal concerns in the schools they teach and with the district.

As cultural brokers, Aboriginal teachers experience the pressure to educate their colleagues on very basic to complex issues facing Aboriginal people, past and present. Often the Aboriginal teacher is seen as the go-between for teachers to parents and community, even if the Aboriginal teacher is not from the particular traditional community she or he teaches in. For example, in my graduate research I interviewed an Aboriginal teacher who described how she had been called on by a school to speak to the family of a young student who had head lice. I could not help but wonder, if a child of Indo-Canadian heritage had a case of head lice, would the administrator call upon an Indo-Canadian teacher to address the child and family?

The Aboriginal teacher is often looked upon as less qualified to teach. To be hired in many districts, she or he must take jobs that are targeted from funds designated as Aboriginal monies. Voluntarily or not, Aboriginal teachers often become advocates for Aboriginal children, running the risk of being ghettoized as the "Aboriginal expert." Aboriginal teachers often feel that as the brokers to Aboriginal culture and community they are on display, taken out of the cupboard for show—and returned once their usefulness as showpieces has been fulfilled. We often have great debates comparing colonialism and globalization, seeing in globalization another form of colonization. The pressures on Aboriginal teachers are enormous, and it is has been difficult to find non-Aboriginal teachers who are willing and able to share the load.

The position I hold in the BCTF makes me a cultural broker for Aboriginal teachers. This puts someone in my role in a vulnerable position within a union. I must not misplace and or lose touch with my committee and their desires. A person in this position must be here for them—not for personal advancement—and I have to remember this. What is the purpose of my being here if I am not going to further the direction of Aboriginal education?

The role of "cultural broker" is an awkward position that creates great tensions. Without the BCTF's commitment to social justice, my position would not exist. Yet, those who do what is often thought of as the "special" work of social justice, even in a union that defines itself as a "social justice union" are susceptible to the feeling that they have stifled or squandered their voice. They must balance their affiliation and commitment to a social justice union, which like all unions includes members who do not support the work of social justice, with deep attachments to people and struggles that have been marginalized by institutions, including trade unions themselves.

CHAPTER 19

South African Teachers and Social Movements
Old and New

Shermain Mannah and Jon Lewis

South Africa has provided fertile soil for the growth of social movements—
even before the term was officially coined in the present era. Colonialism
and military conquest by the Dutch, British, and Boers was met by stiff
resistance by African societies for over two hundred years. The period of formal
military resistance came to an end with the defeat of the Bombatha rebellion in
1906. But even during this period there were also movements throughout the
twentieth century: through formal organizations—principally the national liberation movement led by the African National Congress (ANC) and trade
unions—but also through social movements, sometimes spontaneous but often
linked to the national liberation movement.

With the banning of political organizations in 1960 and intensified state
repression, formal resistance was driven underground or into exile for over a
decade. The growth of black consciousness beginning in the late 1960s, the
rebirth of trade unionism out of the Durban strikes of the early 1970s, the
Soweto students' revolt of 1976, and the open reemergence of the Congress
tradition set the scene for a qualitative and quantitative leap forward in the
1980s in terms of formal political and trade union organization, increasingly
linked to popular movements in the townships and schools.

And this is really where our story begins. We want to show how
SADTU—the South African Democratic Teachers Union—emerged out of
the political and social movements of the 1980s and how its subsequent
development was shaped by these struggles and the political alliances they

entailed. We will also consider SADTU's response to the "new" social movements that emerged after 1994.

1980s: Social Movement Unionism and the Mass Democratic Movement

In order to understand the formation and subsequent development of the South African Democratic Teachers Union (SADTU)—representing the majority of South African teachers—one has to look to the conditions of the 1980s out of which the Union was born. In particular, the future trajectory of teacher unionism was formed by two great social movements—the trade union movement (principally the Congress of South African Trade Unions [COSATU] formed in 1985) and the education mass movement led by the National Education Coordinating Council (NECC). These formed important structures within the much broader mass democratic movement incorporating youth and civic movements, as well as political formations—especially the United Democratic Front (UDF), widely seen to be the internal wing of the ANC. Millions of the populace were mobilized in a massive and ongoing confrontation with the apartheid state.

The period was characterized by political strikes, mass "stay-aways" (work stoppages organized across unions), consumer boycotts, transportation boycotts, rent and service charge boycotts, and boycotts of apartheid institutions. The period was also marked by an explosion of mass-based popular organization—Parent Teacher Student Associations (PTSAs) in the schools, civic and street committees in the townships, shop stewards' councils that straddled workplace and community, and the student and youth organizations—the "young lions"—all of which rendered the townships "ungovernable." It was out of this maelstrom that SADTU emerged in the late 1980s.

Social Movement Unionism

The term "social movement unionism" was coined by labor analysts in the 1980s to describe similar movements—such as Solidarity in Poland, the Workers Party in Brazil, and COSATU in South Africa—where trade unionism had extended its influence beyond the workplace into the community and the national political arena.

In the case of South Africa, the rebirth of trade unionism in the early 1970s—consolidated with the formation of Federation of South African Trade Unions (FOSATU) in 1979—was initially marked by a fairly narrow concentration on workplace organization, what came to be known as "workerism."

The argument put forward was that it would be suicidal for open trade union organizations to try to confront a repressive state—they should rather concentrate on building membership, training shop stewards, and negotiating recognition agreements factory by factory. In the conditions of the 1970s this made some sense. But with the explosion of resistance in the 1980s and the subsequent increase in violent repression in the townships, the unions could no longer absent themselves from the national liberation struggle. In the South African context social movement unionism came to signify the following:

- Worker control of the unions through elected worker leadership at every level, the role of workplace-based shop stewards being of particular importance
- The development of shop stewards' councils that extended activities beyond the individual workplace
- High levels of participation and democratic control through regular mandating and reports
- A high priority to worker education and cultural expression
- Alliances with community and student organizations and growing willingness to take militant action over repression and national political issues

The conditions of the 1980s also helped forge a wider unity among labor. FOSATU and the resurgent congress-aligned unions, together with the mighty National Union of Mineworkers (NUM), joined together to form COSATU in 1985. Unfortunately, the black consciousness–aligned unions remained outside the new federation. COSATU's subsequent organizational growth was accompanied by much political debate, and eventually the decision was taken to adopt the program of the Congress movement. Subsequently, COSATU was to play a leading role in the internal anti-apartheid struggle, culminating in the unbanning of political organizations in the 1990s and the negotiated transition to democracy in 1994. COSATU's political direction was formalized in the Tripartite Alliance of COSATU and the South African Communist Party (SACP) under the leadership of the ANC.

COSATU itself did not use the terminology of social movement unionism, preferring the term "revolutionary trade unionism." This incorporated the characteristics of social movement unionism listed above, as well as an insistence on the following:

- A refusal to be confined to workplace issues
- The need to develop an independent working class political program

- Within the Alliance a commitment to fighting and consolidating the national democratic revolution (NDR)—to consolidate democracy, to create a nonracial and nonsexist South Africa and to address the massive social and economic inequalities of the past
- Fundamental economic transformation of the socialist nature needed to attain the goals of the NDR

This, then, is the political and trade union tradition that SADTU inherited. The union was bound to respond to the new social movements of the late 1990s within the parameters set by the traditions of COSATU and the alliance.

People's Education for People's Power: The National Education Crisis Committee (NECC)

Although students in the early 1970s sporadically protested against inferior Bantu education for black African students, it was the June 16, 1976, student uprising in Soweto, Johannesburg, that provided the catalyst for increased community and organizational involvement in education provision for black students in South Africa. This was further expanded in the 1980s with the convergence between struggles in the workplace and community-based struggles around working and living conditions and the demand for representative structures in educational institutions (Vally and Motala, cited in Kallaway 2002).

Out of these struggles emerged the concept of "people's education for people's power" that was closely aligned to the Freirian concepts of education as a contested terrain where power and politics are given a fundamental expression. Education represents in Freire's view both a struggle for meaning and a struggle over power relations and dominance. This resonated with the South African activists' vision for an alternative education, an "Education for liberation." The call by the United Democratic Front in the early 1980s to make South Africa ungovernable and accelerate the demise of the apartheid state had a particular resonance with black students who were struggling against the imposition of an inferior apartheid education system. Thus, schools in the townships became the battlefield of the struggle, and students, parents and communities united under the slogan of "liberation first, education later."

People's education had varied definitions: an education movement, a vehicle for political mobilization, an alternative philosophy of education, or a combination of all three (Motala and Vally 2002). Some of the key features of "people's education were these:

- Based on decades of education resistance, "people's education" was a rejection of apartheid education, which is education for domination.
- It had an underlying assumption that education and politics are linked and, consequently, that the struggle for an alternative education system could not be separated from the struggle for a nonracial, democratic South Africa.
- "People's education for people's power" was then at the same time an educational strategy and a political strategy. Through "people's education," people would be mobilized and organized toward the goal of a nonracial democratic South Africa; but at the same time through People's Education, people were beginning to develop a future education system.
- Central to the success of "people's education" was organization of all sectors of the people to take control of education and their lives. Students, teachers, and parents needed to build democratic organization in their own sectors, as well as establish strong working alliances and mutual understanding.
- "People's education" as an education system would be controlled by and advance the interests of the mass of the people.
- Arising out of the education crisis, "people's education" initially addressed itself to formal, school-based education. People's Education was intended to educate and empower all, not only school students.
- It would instill democratic values such as cooperative work and active participation—in opposition to current authoritarian and individualistic values dominant in schools.
- It would stimulate creativity and critical thinking to equip students for the future.
- Educational practices implementing the principles had to be developed, particularly with teachers.

Kallaway (2002) argues that the centrality of educational discourses and practices in establishing a new language of resistance and democracy in the form of people's education was a necessary condition for the political transition of the 1990s. Thus, the concept of people's education was seen as resonating strongly with the aspirations of working-class people, who viewed it as a movement that would lead to their economic, political, and social emancipation. Thulas Nxesi (2006), explains:

Workers in the factories, youth in the townships, students in schools and education institutions, mass and underground activists, radical intellectuals, cadres of Umkhonto we Sizwe, militants at all levels were seeking answers to the pressing strategic, tactical and organizational questions of the day. . . . Increasing numbers of our people understood the essence of Lenin's political maxim: Without revolutionary theory, there can be no real revolutionary movement. Its aim was to lead the working class towards the strategic goal of establishing a socialist republic and the more immediate aim of winning the objectives of the National Democratic Revolution which is inseparably linked to it.

In line with the Freirian notion that education includes and moves beyond the notion of schooling, student boycotts in the 1980s took on a new significance when they became linked to demands and campaigns around community and shop floor issues. Motala and Vally (2002) contend that this collective action with students and the broader community revealed that grassroots intervention in the education arena was capable of having a powerful impact on events in civil society far beyond the realm of the schools. The emergence of the National Education Crisis Committee (NECC) marked a shift of oppositional strategy in education from simple school boycott to the construction of alternatives, two processes that were viewed as having a dialectical relationship.

The National Education Crisis Committee (NECC—later renamed the National Education Coordinating Committee) emerged in 1985 as a direct result of this crisis in education. The NECC, a grassroots social movement, brought together over two hundred community, youth, educational, trade union, religious, and other structures to form a broad social movement that worked tirelessly to provide for young people ejected from the apartheid state's formal education system due to political resistance, as well as to explore alternatives to apartheid education. Its formation provided the opportunity for the ideas of people's education to take shape. The path aspired to by the NECC resonated strongly with the call for free and equal public education.

The unbanning of political organizations in the early 1990s and the beginning of negotiations toward a new political dispensation saw a shift in the role of the NECC as the leading agent for education transformation, with the ANC increasingly taking over responsibility. Unlike the grassroots community mobilization strategy of the NECC, the ANC's policy development emphasis was on macro and systemic change. Hence, the role of communities and civil society groupings became increasingly marginal to the overall education transformation project. Motala and Vally (2002) argue that this resulted in the discourse and content of education policy shifting from radical demands of people's education, which focused on social engagement and democratizing power relations, to a discourse emphasizing performance, outcomes, cost effectiveness, and economic competitiveness. This signaled a move toward neoliberal principles underpinning education transformation in South Africa.

SADTU's emergence in the 1990s was accompanied by a lull in active community mass participation in education transformation. Instead, education policy development became a contested consultative process with major stakeholders including trade unions, NGOs, and the new democratically elected government. The NECC was dissolved and the National Education Training Forums (NETF) was established in the early 1990s as a formal platform for state, civil society, and other stakeholders to engage in education policy transformation. However, Motala and Vally (2002, 186) argue that the NETF, "while framing its goals in terms of the principals of democracy, inclusiveness, and transparency, in fact reflected the interests of its broad stakeholder grouping which favored business and the state." The result was a decline in grassroots-based community participation in education transformation that favored social justice and redress to an overreliance on technocrats who promoted an education system that related to issues of economic growth and human capital and resource development (Kallaway 2002).

Despite the limited success of the proponents of "people's education", the vitality of the movement in mobilizing all sectors of the community provides a real example of a challenge to apartheid education and the development of an alternative education that resonated with the will and vision of oppressed communities in South Africa. Motala and Vally (2002, 189) state that "despite the obvious discontinuities, the vitality of PE embodied in student organizations, teacher unions, NGOs and potentially, in the organization of parents and governing bodies, creates possibilities for challenging the dominant trajectory and thereby continuing a rich legacy."

Teacher Unity: The Formation of SADTU

As Harold Samuel explains in the essay he has contributed to this book, teacher unionism historically was divided along racial, ethnic, geographical, and political lines. In looking for the origins of SADTU we must go back to the mid 1980s—at the height of the struggle against apartheid—when progressive teacher organizations were meeting to strategize on how to unite and pull in teachers organized in the established teacher organizations. Only partial unity was achieved when SADTU was launched in 1990 with 30,000 members. However, SADTU went on to become by far the largest teacher union, with a membership today of 230,000, representing nearly two-thirds of teachers in South Africa.

As a new organization struggling to establish itself in the 1990s, SADTU had a number of priorities.

1. The Struggle for Recognition 1990–93

In the early days, under the old regime, SADTU was engaged in a mammoth battle to secure its very existence. It was a time of strikes over recognition and against victimization. By 1993 certain basic rights for the union and the members were secured:

- The right to organize
- The right to form and join the union of choice
- The right to fair procedures
- The beginnings of the right to negotiate

However, this was still far short of the full labor rights enjoyed by most workers in the private sector.

2. Political Identity: Affiliation to COSATU

1993 was a turning point in South African labor history. SADTU's affiliation to COSATU meant that teachers were now part of the broader working-class movement in the country. It marked an important step away from the traditional divide between white-collar and blue-collar workers.

The struggle for full labor rights now began in earnest. As part of COSATU, and alongside the other public sector unions, SADTU now campaigned for the Labor Relations Act (LRA), which would cover all workers. As a result of COSATU's LRA Campaign and with the election of a democratic government in 1994, the new LRA was finally enacted to extend full labor rights to public service workers. This allowed SADTU to better safeguard the salaries of members.

3. Professional Issues

SADTU was established on the basis of the need to marry trade unionism with professionalism. Professional issues were raised from the start. Following major successes on the collective bargaining and political fronts, SADTU's 1995 National Congress resolved to redouble its efforts toward professional development and education policy work and to devote resources to this end. The objective was to ensure that the Union played a leading role in the policy debates of the time and in empowering teachers to cope with the new demands of education transformation.

In other words, when the new social movements arrived on the scene in the late 1990s, SADTU had its hands full. Politically it was part of COSATU and the Alliance and still struggling to find its feet. On the labor front, it was engaged in a massive recruitment campaign and battling to protect wages, which was increasingly difficult after 1996 with the adoption by government of a conservative economic policy, Growth Employment and Redistribution (GEAR). On the education policy front SADTU was the major stakeholder in the process of education restructuring.

Post-1994: The Alliance, GEAR, and the New Social Movements

The initial hopes and expectations generated by the 1994 elections began to dissipate in the late 1990s as many citizens grew frustrated with the lack of delivery by the new government and the increase in unemployment and pervasiveness of poverty. Thus there was an emergence of social mobilization around issues of socioeconomic rights and delivery of services like housing, access to antiretroviral treatment, and water and electricity and in opposition to aspects of government policy—particularly macroeconomic policy—the so-called GEAR strategy—which included reduced public spending, "right-sizing" of the public sector, conservative monetary policy, and privatization. Opposition took a number of forms, including the outright rejection of government strategy and policy by the new social movements, growing opposition from within the Alliance led by COSATU and later joined by the SACP, and emergence of the Treatment Action Campaign (TAC), an attempt to combine mass mobilization with lobbying within the Alliance. We describe each more fully.

GEAR generated social opposition. Desai (2002, cited in Ballard, Habib, and Valodia 2006) describes how the poor, calling themselves the Concerned Citizens Forum, (CCF), were forced to defend themselves against evictions and disconnections from water and electricity. He argues that their actions are understood as a "local response" to "globalization." However, Dwyer (2006) argues that participants of the CCF cannot be straightforwardly categorized as having a single, coherent political consciousness or identity in opposition to "neoliberalism" and "globalization." He maintains that while drawing on a range of often contradictory histories, discourses, and symbols, these people identify common cause with each other largely through shared experiences. In this way, a perceived failure of delivery that does not match people's expectations provides the potential basis for activism. Although the

CCF has been criticized by the ruling party and others for not engaging with the state in a constructive manner, Dwyer (2006, 104) argues that by invoking constitutional rights and taking the government to court, the CCF has used the political spaces opened up since 1994 to claim their rights, thus contributing to the establishment of liberal democracy.

TAC was formed as a response to the South African government's reluctance to acknowledge the scourge of HIV and AIDS and effectively address the pandemic that was raging through the country. TAC mounted a highly effective campaign focusing on securing treatment for people with HIV and AIDS. Its success is all the more impressive in light of sustained hostility and HIV and AIDS denialism from the highest levels of government. The ability of TAC to create a moral consensus and forge strategic alliances was central to its success as a social movement.

The mobilization of TAC in 2001 against the might of multinational corporations bent on preventing the South African government from importing cheaper generic medicines was a watershed in this country's history of social struggle. The alliances built with COSATU and other international organizations like Medicines sans Frontières and Oxfam created a formidable force that saw workers take to the streets in support of TAC demands. TAC's activist base and COSATU's mass worker membership formed a powerful combination to make grassroots participants aware about HIV and AIDS and pressure the government to sanction a plan to distribute antiretroviral medication to people living with HIV and AIDS. Despite the strength of the tripartite alliance between the ANC, the SACP and COSATU, the COSATU National Congress of 2000 witnessed a major difference of opinion between the federation and the country's president, when the federation challenged the president of the country's questioning of the link between HIV and AIDS.

The success of the TAC campaign rests on the effective way it managed to find a balance between using the courts and using the masses in the streets. The use of anti-apartheid–type tactics of international solidarity, broad alliances, and civil disobedience proved useful even during the current dispensation. However, unlike under apartheid, TAC was challenging a legitimate government elected by the people of the country. In the current democratic South Africa, winning and retaining public opinion is of greater concern now than during the anti-apartheid struggle, when support was usually taken as a given. Friedman and Mottiar (quoted in Habib et al., 2006, 25) warn that "the legitimacy of the government and the popularity of the ruling party are new realities that activists forget at their peril: 'A major tactical error would be to lose support among our members as other social movements have done when they are seen to be threatening democratically elected leaders' (Achmat 2004)." Unlike many other social movements that have run afoul of the current government, TAC has been very strategic in appreciating and accommodating formal democracy in its engagement with government. This tactic has also created an enabling environment for the continued support of COSATU.

The real test of the alliance with COSATU emerged when the TAC went on a civil disobedience campaign in response to government's failure to sign an agreement consenting to an AIDS treatment plan. COSATU was unwilling to participate, fearing participation would be perceived as a rebellion against the government. Friedman and Mottiar (2006, 31) argue that alliance politics is not simply a matter of gratefully accepting the support of those who agree, it also assumes that common ground can be found with those who differ. They maintain that it also involves compromise. Concessions were made by the TAC to retain COSATU as an ally, and it was agreed that COSATU's failure to participate would not jeopardize the alliance (Friedman and Mottiar, t al., 2006, 33). Thus, the TAC was able to work through a strategic challenge in its alliance with COSATU and maintain its autonomy.

Friedman and Mottiar (2006, 38) sum up the reasons for the success of TAC: "The difference between other social movements lies in the reality that the TAC, unlike other social movements, engages with the post-apartheid system and accepts that rights can be won within it. To use the law implies that it is not inherently biased against the poor. To lobby politicians implies that those who demand equity can find allies in mainstream politics. To help the 'roll out,' albeit in a way which may require confrontation, implies that the government can, with prodding, meet the needs of poor people living with HIV/AIDS."

SADTU's Role and Future Challenges

How well has SADTU fared as a "trade-union social movement" under the current democratic government? The SADTU General Secretary explained three key challenges, at the sixth national congress in September 2006:

- The theme of this congress, "*Empowering educators to reposition SADTU for people's education and working class power,*" I believe, raises the issues in a clear manner: unless we place the interests of the working class and the poor at the center of our NDR—and education transformation—we are headed down the same road of betrayal as most of the postcolonial states. Other class forces are already laying claim to the fruits of victory against apartheid. The alliance between white capital and the new black bourgeoisie seeks to ward off the demands of the working class and the poor to protect their selfish interests.

- Last year we celebrated fifty years of the Freedom Charter. Drawn up in the darkest days of our struggle by the Congress of the People, thousands of delegates meeting down the road in Kliptown, that document is as radical and clear in its vision today as it was all those years ago. In relation to education, the Freedom Charter proclaimed: "The doors of learning and culture shall be opened to all." It declared the goal that "Education shall be free, compulsory, universal, and equal for all children." Are we still on course to achieve this?—We have to ask ourselves, how far have we managed to open those doors? In the former model C schools and the universities, I would suggest, those doors remain firmly closed to the mass of working-class and poor people in this country.

- We have identified neoliberal economic policies as one of the challenges to education transformation. In fact, the worst privations of South Africa's neoliberal GEAR policy had come to an end by the time of our last congress in 2002. There have been real increases in education budgets since then, although not sufficient to address the massive backlogs caused by Bantu education. To save jobs we had to accept the rationalization and redeployment policy—which was highly disruptive to the lives of educators. The scale of restructuring in the public service was limited by the timely intervention of COSATU. Our last congress also identified problems of equity and access due to the funding system, and here change has been very slow in coming. As congress you need to direct us as to the way forward in this respect.

Mr. Nxesi provides a synopsis of some of the challenges facing the union. However, the union's response to these challenges cannot be analyzed without first understanding the political context in which the union has been operating. Although SADTU remains committed to its policies for education and societal transformation, the context in which it is forced to exist has changed. Most significant in this regard have been the position that SADTU occupies in the ANC Alliance as an affiliate of COSATU and the changing macroeconomic context and policy positions of the ANC government. The union's inclusion in the processes of policy formulation without the ability to change the overarching macroeconomic policy framework governing them has prevented SADTU from playing its intended role in policy formulation and as vanguard of the poor. Hence, SADTU has unintentionally been sidelined by the neoliberal logic of GEAR and struggles to deliver on its vision for free quality public education within these neoliberal rationalities. It would be pointless for SADTU to continue to operate in its 1990s style while government has embedded its neoliberal principles in policy and implementation. The spaces for genuine consultation have shrunk since 1998, and it has been essential for SADTU to adjust its advocacy strategies in response to this change in the consultative environment.

Furthermore, given SADTU's position in the education arena, it has failed to change the mindset within the state bureaucracy. It appears that while SADTU has repeatedly committed itself on paper to fighting for a system of free primary and secondary education, it has done little in practice to ensure reshaping of existing policies on school funding. SADTU has supported Outcomes Based Education as a progressive educational framework. However, the system is now fraught with bureaucratic processes of implementation and lack of resource mobilization in certain areas. Hence, teachers and learners have failed to benefit from it due to lack of proper training and support, continuing historical deficiencies, and an overemphasis on assessment and administration.

Conclusion

We have argued that SADTU's response to the new social movements of the late 1990s was constrained by its own history:

- Its emergence out of the old social movements of the 1980s—the mass labor, education, community and democratic movements of that time
- Its position as an affiliate of COSATU and consequently a part of the Tripartite Alliance that underpins the ANC in government—which in turn influences the kind of actions Alliance partners can take in relation to their own government
- The quantum growth for SADTU on every front that kept the union's hands full during the 1990s—recruiting members, establishing structures, fighting for recognition, establishing labor rights, improving teachers' conditions, and developing education policy and training capacity
- The rapid growth of the union, which entailed a degree of bureaucratization and also reduced its ability to mobilize for political protest

We have also argued that the strategy and tactics of the majority of the new social movements—outright opposition to the ANC in government—also militated against cooperation with SADTU and other Alliance partners.

This strategy also served the new social movements ill, since they failed to grow or to pose a significant challenge to existing power relations. The exception, of course, is TAC, which adopted a dual strategy of engaging with government and its allies alongside mobilizing mass protest. In the case of TAC, SADTU and COSATU have worked very closely with them.

While history has shaped the response of SADTU and the labor movement to the new social movements, they are not simply prisoners of that history. The labor movement's response to changing conditions of globalization is also evolving. Several factors are relevant. First, the spirit of the 1980s—the heyday of the old social movements—may be dormant, but it is not dead. True the youth and student movements have been largely demobilized, although the formal organizations—ANC-YL, COSAS and SASCO—remain, and indeed have been reinvigorated in recent years with the launch of the rapidly growing Young Communist League (YCL). The union movement in particular—while it often felt marginalized within the Alliance never actually suffered any major defeat and continued to grow in numbers.

Also, the "left" within the Alliance—the majority of COSATU and the SACP—while constrained by the alliance with the governing party, have opposed policies that were not in the interests of the poor and the working class. On macroeconomic policy the federation has led the opposition to government and mounted a series of national strikes since 1999 to oppose privatization, poverty, and unemployment. In addition, shifts within the Alliance and growing class conflict between the new black bourgeoisie and the left have created tensions within the Alliance and the ANC, which are reflected in the current leadership contest and the mass movement around Deputy President of the ANC, Jacob Zuma.

There is also an ideological dimension to the growing tensions within the Alliance. The national liberation struggle was fought under the leadership of the ANC with an eye to building and maintaining maximum unity. But there was also the understanding within the ranks of the national liberation movement that South Africans were involved in a two-stage revolution: crudely put—first political liberation, and then class struggle for economic emancipation and socialism. At least in part, debates within the Alliance refer to differences of tactics and analysis as to whether we need to continue to stress the need for maximum unity at this stage of the NDR, or whether we are experiencing intensified class struggle within the Alliance.

In terms of the wide political context, the balance of forces appears to have moved in the direction of progressive forces. On the world stage U.S. imperialism is losing ground: in the so-called war on terror, in Latin America, and with the rise of China and India. Internationally and nationally, neoliberal economic policies have not delivered. In South Africa, the major challenges are poverty and inequality. The social and economic conditions that underpinned the social movements of the 1980s and 1990s remain.

Finally, the purely oppositional tactics of the new social movements proved ineffective—TAC being the exception here. This suggests two things: that the struggle against neoliberalism and globalization will be fought out largely within the Alliance, with SADTU playing a role in this, and that the example of TAC provides a model for the future to revitalize social movements both old and new. In relation to this last point we should note the establishment three years ago of the Education Alliance consisting of SADTU, COSAS, SASCO, ANC-YL, YCL, and NASGB. The fledgling organization brings together organized teachers, students and youth, and to some extent parents represented in SGBs—these are the same groups that underpinned the NECC—united around the demand for free and equal quality public education. While the Education Alliance has established a national voice, it remains to be seen whether this can be translated into mass mobilization and pressure for education transformation.

References

Ballard, R.; Habib, A.; Valodia, I (Ed) 2006 *Voices of Protest: Social Movements in post-apartheid South Africa*. University of Kwa Zulu Natal Press, South Africa

Dwyer, P. 2006, *The Concerned Citizens Forum-A fight within a fight* in Ballard, R.; Habib, A.; Valodia, I (Ed) *Voices of Protest: Social Movements in post- apartheid South Africa*. University of Kwa Zulu Natal Press, South Africa

Friedman, S. and Mottiar, S. 2006 *Seeking the high ground: The Treatment Action Campaign and the politics of morality* in Ballard, R.; Habib, A.; Valodia, I (Ed) *Voices of Protest: Social Movements in post-apartheid South Africa.*" University of Kwa Zulu Natal Press, South Africa

Kallaway, P. (Ed) 2002 *The History of Education under apartheid 1948–1994. The doors of learning and culture shall be opened*. Maskew Miller Longman, South Africa.

Motala, S. and Vally, S. 2002. *People's Education: From People's power to Tirisano* in Kallaway, P. (Ed) *The History of Education under apartheid 1948–1994. The doors of learning and culture shall be opened*. Maskew Miller Longman, South Africa.

Naledi 2006. *SADTU's impact on education policy* Report to SADTU 6th Congress–2006, SADTU

Nxesi, T. W. 2006, *Summary of the Secretariat Report*, SADTU 6th National Congress–Midrand, 30 August-2 September 2006, SADTU, South Africa

Oldfield, S. and Stokke, K. 2006, *Building unity in diversity: Social Movement Activism in the Western Cape Anti-Eviction Campaign* in Ballard, R.; Habib, A.; Valodia, I (Ed) *Voices of Protest: Social Movements in post-apartheid South Africa*. University of Kwa Zulu Natal Press, South Africa.

CHAPTER 20

Schooling and Class in Germany
An Interview with Eberhard Brandt and Susanne Gondermann

Mary Compton

(Editors' note: Germany is a federal country, so education systems vary from state to state, as they do in the United States. Most states in Germany have a "tripartite" system, which separates children, usually at the age of ten, into one of three options: main schools, secondary schools, and grammar schools. Each has its own course of study and leaving or exit exam. Students leave the main schools at fifteen, the secondary schools at sixteen, and grammar schools at eighteen or nineteen. If they wish to attend higher education, they must take an exam called the Abitur. A few comprehensive schools, which were the norm in the old East Germany and in some traditionally Social Democratic states like Hamburg and Hessen, still exist, but these are increasingly under attack.

In the PISA study comparing educational achievement in different "developed" countries, described in Larry Kuehn's essay, Germany did particularly badly. The director of the study, Andreas Schleicher, himself a German, has argued strongly this is because of the "tripartite" system, which results in an inferior education for students in the main schools, who are not expected to attend the university.

Germany's teacher union, the Gewerkschaft für Erziehung und Wissenschaft (GEW), is also organized on a federal model. Mary Compton interviewed two GEW officers, Eberhard Brandt, an official in Lower Saxony, and Susanne Gondermann, an official from Hamburg, about class stratification in German schools, and the GEW's response to it when they attended a conference of the National Union of Teachers

(NUT) in the UK in 1986 and has translated their comments from German into English.)

MC: What effect does the tripartite system have on immigrant children or children with special needs?

SG: Children from immigrant families are nearly all in main schools, some in comprehensives, and very few in grammar schools. Only if their parents come from embassies or from a middle-class background are they more frequently at a secondary or grammar school. But most children from an immigrant background are very strongly disadvantaged by the tripartite system.

MC: Has the poor showing of Germany in the PISA study meant that the tripartite system is being challenged?

EB: Only by the GEW—which says that this system achieves the most extreme segregation effect in the world, but the education politicians are arguing exactly the opposite. Everywhere we have right-wing governments of Christian Democrats who base themselves on the tripartite system and who rely on the old moth-eaten theories of "giftedness" and say there must be schools for those who have "practical" gifts and who mustn't be overburdened with scientific theory and who must do practical subjects in school and then they will be happier. "We must call the 'bad' students 'bad,'"—as the former federal president Roman Herzog said—"and the elite must be allowed to be elite and more exclusively together." We can only mourn such sentiments.

MC: So the old divisions are as strong as ever?

EB: No. Pupils are even more strongly separated than they were because the curricula are being rewritten for the main schools.

SG: For example, languages—they only learn one and at a very low level, and as for science—it's not so much science as nature study, so the pupils have much lower expectations. This means that pupils who have attended a main school can't train for lots of the jobs, which are taught at further education colleges, because they haven't got the basics—in math, physics, or chemistry, for example

MC: And they leave the main schools at fifteen, don't they?

EB: They leave after nine years of education, but because we have the system in Germany of repeating years if end-of-year marks aren't good enough, some pupils who have had to repeat years can leave school after only the seventh or eighth grade. Ten percent of pupils in different federal states have no school leaving exam of any sort.

MC: And what facilities are there for children with behavioral problems or learning difficulties?

EB: There are special schools The children are sorted out in primary schools and sent to three or four different sorts of special schools—schools for children with physical disabilities, or learning difficulties—they are separated

MC: And this tripartite system fits very well with the neoliberal agenda doesn't it, where they want to separate children between those who are cut out to be part of an elite managerial class and those who will become "flexible" manual workers?

SG: Yes it fits—but it isn't based on that in Germany. The division based on social class has never really been overcome. It was the Weimar republic in 1920 that first brought in four years of primary school for all, but after the Second World War the opportunity was missed to bring in integrated secondary schooling for all—except, of course, in the old German Democratic Republic.

MC: Does that mean the neoliberal agenda is completed in Germany?

SG: No. Contemporary politics is adding to it—it is sharpening the segregation that we already have—particularly insofar as those few schools that are integrated are being damaged by neoliberal policies, for example, through national testing and leaving exams and so-called educational standards. These effects restrict pedagogical freedom in schools and mean that those reforms that have already been achieved in comprehensive schools are set back. For example, in school years nine and ten, where there used to be lots of project work, teachers can't do that any more but have to concentrate on the leaving exams.

MC: And is privatization also an issue in Germany?

EB: Not directly privatization. We have very few private schools, and that hasn't changed, but there are new forms that could represent a transition to privatization. For example, the new "standards" and inspection are supposed to show that some schools don't reach the required standards, whereas others look very good. This mechanism strengthens the segregation effect even further. The second thing is that schools are treated like businesses and given budgets so they are almost like private firms. Then new laws in some states— for example, in Lower Saxony—have brought in foundation schools, where the foundations decide the school program but the state pays the teachers and the community pays for the school. The foundations recruit staff and can decide, for example, only to have Catholic teachers.

MC: Are these foundations mostly religious?

EB: Churches are certainly very interested—both Catholic and Protestant. They might control a grammar school, for example, and then join it with a primary school, and this can have serious implications for the primary schools as well, which have always been integrated up to now. This isn't directly privatization, but it's the influence of private interests.

MC: Can private firms do this as in England, or is it different?

EB: In theory any private individual could do it, and there is only very limited state control over these schools. They can even change the state curriculum, although the degree of freedom to do this is not yet clearly defined—and they don't have to do the state leaving exam. These are mostly grammar schools, and they don't have to do the central Abitur—they can make up their own exam and just have it checked by the education ministry.

SG: In the further education colleges in some states, the representatives of business have got more influence on the governing bodies. This is relatively new, and it's a significant change. Also, for example, the marking of tests is often done by private firms—who make money out of it, of course. And as state provision declines, more and more professional development is coming from private firms and is incidentally very expensive

EB: And in Lower Saxony they are thinking of outsourcing the educational psychology and legal departments of the school system who were responsible for defending teachers against complaints from parents, for example

MC: How does the GEW fight against these developments?

EB: Up to now the GEW has not had a unified position and plan of action. In some states where the policies have gone a long way the GEW, is keeping up. In others these issues haven't really been tackled. At the last union conference we passed a resolution on the policy of so-called self-managed schools—the equivalent of the English locally managed schools—where we analyzed and described this development using experiences from England to inform our analysis, and we voted to fight against it and instead to fight for an integrated, democratic school for all pupils. But you can't get that until you have absolutely finished off the government policy of self-managed schools. Yet there is a position in the GEW among some officials, who say that self-managed schools are really a great opportunity to fight for our positions—that we can introduce our educational ideas into such schools—and that we shouldn't be so timorous. We have to continue this discussion and complete it

MC: Perhaps this is rather a naïve position—in England we've had local management of schools for a long time—the problem is always with funding—there is never enough money and the head teacher becomes more like a manager than a teacher. And now they're saying that perhaps we don't need a teacher anymore to be a head teacher—we can just use a businessman.

SG: In the GEW this question of privatization and of the new management of schools still hasn't really been discussed through. This new management system for schools is very damaging for schools. Treating schools like businesses, the attack on the curriculum, control from above—these are all very dangerous developments and must be rejected. It's wrong to characterize this as privatization although it might lead the way to it, and it has no place in schools.

EB: It is essential that we explain how this new form of management works. For example, in our schools managers are having to agree on targets with the authorities, and the schools have to compete one against the other. We have to show the way children with problems are sorted out and how the climate is changing in the staff rooms. Under the promise of more freedom there is more authoritarianism.

SG: It's not only more authoritarianism. Also the political influence on the curriculum and the political segregation of pupils is very dangerous in a conservative state. It makes social segregation even starker. And working conditions are worsened because of reduced staff budgets. We really need to recognize all this together as a union. In order to fight this we have to work together with parents' organizations and pupils' organizations against central testing. I think it's gone so far in Germany that the only thing which will help is an agreement to have an effective national boycott of tests

MC: Would parents agree with that?

SG: In Hamburg we have begun the debate. We've got a parents' group with parents' organizations, the comprehensive school parents and the parents organized by the social democrats, for instance, who have built a campaigning organization called Protest Gegen Tests and who have called for the boycott of tests. They haven't been given enough support by the unions—they have to collect money from individual unionists so that speakers can be invited. We have missed a lot—we haven't analyzed the international developments, and we haven't done enough work in the GEW working through the disagreements and the influence of political parties, which of course have an effect on the GEW.

EB: In Lower Saxony we've had lots of meetings where speakers have talked about self-managed schools and privatization—there has never been such a big demand, and we try to pass resolutions in all the schools and send them to the state governments

MC: Is that organized by the GEW?

EB: Yes, but all the teachers are invited, and they are well attended. We put the educational and the workplace interests together because more and more duties of teachers are taken away from teachers and given to unqualified people; for example, mentoring, which is done by badly paid assistants. And also unemployed people are forced to do work for one euro an hour doing work that was previously done by teachers, for example, running clubs in the afternoon, sport, supervising children doing homework—and they have no training at all. I was in a big comprehensive school recently where they had twenty-five assistants, and the budget had been so severely cut that they could only keep the school going through using unqualified people—otherwise it would have gone broke. This of course intensifies the work of teachers, who can lose clubs and sports practices as part of their contractual duties and are only left with teaching the main subjects and doing even more marking and preparation.

SG: And we have to do something very complicated as the GEW—we have to correct a mistaken strategy. This resulted from colleagues only seeing a part of what was happening—the material worsening of their salaries and conditions, and then they said if the government won't sort that out, then we'll do without integration; we won't go on class trips because it's extra work. That is a big problem because it makes our work worse because it militates against our educational aims and it can destroy cooperation with parents and pupils.

MC: So you're saying this was the wrong kind of industrial action and that a test boycott for instance would be better?

SG: Exactly

MC: How big a difference could EI make to your struggles in Germany?

EB: I think these issues must become the center of attention for EI—so that we can share experiences and find out what is happening in other countries and what opposition strategies have been tried and are possible. Here at the NUT [conference] we have met colleagues from New Zealand, Australia, Israel, the United States, and Canada, and they have all seen the same developments, and they say we are all working in too isolated a way.

CHAPTER 21

Education or Mind Infection?

Nurit Peled-Elhanan

(Editors' note: Throughout the world teachers play a prominent part in social movements for peace, social justice, and human freedom. No book about the global assault on teachers and teaching would be complete without mention of connections between educational change and political movements to bring justice and equality to the world. Nowhere is the relation between schooling and social change more pressing than in the Middle East—which is a primary location of the struggle for dominance over resources and markets being pursued by neoconservatives in the United States and their allies internationally. One result of this ongoing conflict is that millions of children in the region are not receiving a proper education, and often no education at all. Schools are shuttered, sometimes temporarily and often permanently, due to violence, terrorism, military conflict or the poverty and privation that accompany deadly struggles.

Nurit Peled-Elhanan is an Israeli academic and educationalist who has written widely about the anti-Palestinian bias in Israeli textbooks. Sadly, she is also the mother of a little girl who was killed by a Palestinian suicide bomber. In her speeches and writings she advocates for the rights of the Palestinians and for an end to the illegal occupation of the Palestinian territories. She is the 2001 laureate of the Sakharov Prize for Human Rights and the Freedom of Speech awarded by the European Parliament.)

This chapter is dedicated to my friends Salwa and Bassam Aramin, whose little daughter, Abir, was shot in her head at the age of ten by an Israeli soldier who will never be brought to trial. Bassam is one of the founders of the Combatants for Peace movement, of which two of my sons, Elik and Guy, are members.

Vignette 21.1

Below are excerpts from a Human Rights Watch report "Second Class: Discrimination Against Palestinian Arab Children in Israel's Schools" (www.hrw.org), which synthesized data from an investigation of twenty-six Arab and Jewish schools and nationwide statistics compiled by the Israeli government. Nearly one-quarter of Israel's 1.6 million schoolchildren are Palestinian Arab citizens and are educated in schools run by the Israeli government but operated separately from those of the Jewish majority.

The report found striking differences in virtually every aspect of the education system. The Education Ministry does not allocate as much money per head for Palestinian Arab children as it does for Jewish children. Their classes are 20 percent larger on average. They get far fewer enrichment and remedial programs—even though they need them more—in part because the Ministry uses a different scale to assess need for Jewish children. Their school buildings are in worse condition, and many communities lack kindergartens for three- and four-year-olds. Palestinian Arab schoolchildren do not have the same access to counseling and vocational programs. One of the largest gaps is in special education, where disabled Palestinian Arab children get less funding and fewer services, have limited access to special schools, and lack appropriate curricula. Arabic is an official language and the language of instruction in Israel's Arab schools. Nevertheless, the report concluded that the Israeli government devotes inadequate resources toward developing Arabic curricula in general, and Palestinian Arab teachers have far fewer textbooks and teaching materials at their disposal than their Jewish counterparts. The report observed that some of the content, especially the mandatory study of Jewish religious texts, alienates students and teachers alike.

"Government-run Arab schools are a world apart from government-run Jewish schools," said Zama Coursen-Neff, counsel to the Children's Rights Division of Human Rights Watch. "In virtually every respect, Palestinian Arab children get an education inferior to that of Jewish children, and their relatively poor performance in school reflects this." . . . Palestinian Arab children drop out of school at three times the rate of Jewish children and are less likely to pass the national matriculation exams for a high school diploma. Only a handful make it to university. Among Palestinian Arabs, the Bedouin from the Negev Desert fare the worst in every respect. "Discrimination is cumulative, and at each level, more Palestinian Arab children are winnowed out," Coursen-Neff said . . . "The government has admitted that it spends more per Jewish child, but it hasn't changed its policies," said Coursen-Neff. "The children who need the most—Palestinian Arab and, especially, Negev Bedouin—are getting the least." Last fall the government promised extra money for Arab education. This year it is not delivering on that promise in the 2002 budget.

Salwa and Bassam, like myself and my family, are victims of the cruel occupation of Palestine, which has been corrupting our country for forty years and has turned the Holy Land into a wasteland, a land where so many children are buried, including our two daughters—Smadar (the bud of the fruit) and Abir (the perfume of the flower), murdered at an interval of ten years, ten years during which this land has filled with blood and where the underground kingdom of dead children on which we tread day by day and hour by hour has grown to overflowing.

But children are abstract entities for the military, and grief is a political tool. Living on both sides, that of the victims and that of the killers, I keep asking myself, what are the means by which our good Israeli children are mind-infected to the point of humiliating, torturing, and killing other children, their parents, and their grandparents just because they are neighbors. In the so-called enlightened Western world many people consider themselves very well founded when they blame Islam for suicide bombing and terror. But who would ever blame Judaism for the atrocities committed by Jewish soldiers or for their racist education? And yet, Jewish ultra Orthodox children who have never left Brooklyn know that to kill Arabs is a *mitzva* ("holy commandment") for they are *vilde hayeth* ("wild beasts"). And Israeli secular children, from a very tender age, learn to admire biblical Joshua, the glorified leader who murdered the whole population of Jericho in the name of God. Then they learn to admire the prophet Eliyahu—the Jewish Santa Claus—who killed the 450 priests of Baal because they practiced a different religion.

Neither Judaism nor Islam—nor any religion for that matter—is the cause for murder and terror. Greed, megalomania, and racist education are the cause.

Miseducation about Palestinians

In this chapter I shall be looking at school books, all of which are currently studied in Israeli mainstream secular Jewish schools, authorised by the ministry of education and reflect the national curriculum.

Israeli children who actually perform massacres of "Arabs"—Palestinians or Lebanese—grow up for the most part without ever meeting an "Arab" face to face let alone talking to one. All they know is that there are "Arabs" (never "Palestinians") living in *our* Eretz Israel (not in *their* Palestine), although they have twenty-one other countries to choose from. The "Arab" citizens of the state of Israel are usually defined generically as "the non-Jewish population [or sector]" while the Palestinians living under occupation, deprived of all human rights, are referred to as "Arabs." Who are they?

In the geography schoolbook *Israel—Man and Space* (2002, 12) we find a definition: "The Arab Population: Within this group there are several religious groups and several ethnic groups: Muslims, Christians, Druze, Bedouins and

Circassians. *But since most of them are Arabs they shall be referred to henceforth as Arabs*" (emphasis added).

These "Arabs" are presented in schoolbooks as the uncultivated part of the population, the backward element that impedes Israel's progress. For instance, "The Arab society is traditional and objects to changes by its nature, is reluctant to adopt novelties. . . . Modernization seems dangerous to them . . . They are unwilling to give anything up for the general good" (Aharony and Sagi 2002, 303).

In another Israeli geography textbook, *People In Space* (1998, 76), a graph depicting average marriage age for women as one of the characteristics of development manages to locate Israel as the last, shortest bar in a line of "Developed Countries" thanks to a minuscule footnote: "The graph refers only to the Jewish population" (Rap and Fine 1998, 76).

Apart from being a developmental burden, the Palestinian citizens of the state are presented in the educational discourse, as they are presented in the political discourse, as "a horrifying demographic problem," and a security threat. For instance, in the geography schoolbook *Geography of the Land of Israel* (240), we find an explanation for the need to "Juify" the Galilee: "The purpose of the foundation of the Mitzpim[1] is to preserve the national land and keep it from illegal invasion of non-Jewish population, to purchase land for development in order to prevent a territorial sequence of non-Jewish settlements, for fear that an Arab sequence would cause the separation of the Galilee from the state of Israel."

The Palestinian refugees who live under occupation, deprived of all civil and human rights, are defined in history schoolbooks as a poisoning problem: "Although Israel came victorious out of the survival-war that was forced upon her, *the Palestinian problem would poison* for more than a generation the relationships of Israel with the Arab world and with the international community" (Bar-Navi and Nave 1999, 239; emphasis added).

Professor Bar-Navi explains that annexing the West Bank to the state of Israel would create "an absurd situation where the Jews would be a minority in their own land" (Bar-Navi 1998, 249). This, he adds, would turn "the Zionist dream into a *South African nightmare*." This historian, praised for being progressive and politically correct (Firer and Adwan 2004), wrote these words after the great victory of Nelson Mandela. To him South Africa is still a nightmare for the white population with whom he equates the Israeli Jews.

The nightmares of the Palestinians themselves regarding this possible annexation are never mentioned.

Visual Racist Representation of Palestinians

Palestinians, or rather "Arabs," are never portrayed as modern, productive, individual human beings, only in caricaturist racist icons or in stereotypical blurred images of terrorists, "Oxfam" figures of primitive farmers, and refugees. One never sees their faces, only what Barthes (1977) termed the "object-signs" of their stereotype: traditional dress, Kafieh or scarf, the characteristic camel of the stereotypical Arab men. The women are always squatting or sitting on the ground in what is called "oriental position," surrounded by numerous indistinct children. They are qualified, as third world people are in some European books, as clannish, unwilling to contribute to public good, refusing to become modern. "The[Israeli] Arabs refuse to live in high buildings and insist on living in one-story land-ridden houses…they are unwilling to give up anything for the general good" (Aharony and Sagiog 2002, 303).

Racist Maps

All social categories in Israel are divided into Jewish versus Non-Jewish, configuring "Jewishness" as the norm. For instance, in the textbook *Settlements in Space* (2003, 55) a map depicting rural habitation in Israel is divided into Jewish and Non-Jewish settlements: "Rural habitation in Israel: Blue: Jewish villages; Red: non-Jewish villages"

None of the official maps in Israeli schoolbooks or media present the Green Line, which is the border between the state of Israel and the Palestinian-occupied territories. The Palestinian-occupied territories are depicted as part of the state of Israel, but Palestinian inhabitants, institutions, and sites are missing from maps, photographs, and graphs, as well as from the verbal text. According to Prof. Yoram Bar-Gal,[2] the texts convey the ideological stance of Israeli curriculum planners that man-made borders are an "accidental consequence of cease fire commands which paralyzed military momentum" (Bar-Gal 1993a, 125). Bar Gal maintains that "the borders of Israel as presented on the map represent the right-wing ideological perception which refuses to see the area of the West Bank and Gaza as territory under a different sovereignty" (ibid.). "The educational system continues, therefore, to present maps as a miniature model of reality, and less often emphasizes that this map is a distorted model, which sometimes can 'lie' and contain items that are completely different from reality" (Bar-Gal 1996, 69).

One way of erasing the existence of Palestinians from the land is what is called in cartography "geographic (or topomomyc) silence" (Henrikson 1994). A few examples of this occur in the textbook *Israel—Man and Space*. For instance, in a map of Jerusalem titled "Jerusalem as Capital—Government,

Culture, Administration and National Sites," there are no cultural, administrative or national sites in East Jerusalem except for the Wailing Wall. In a map depicting the distribution of "Arab Population in Israel 2002" (Fine, Segev, and Lavi 2002, 16), Palestinian regions are defined as "areas for which there are no data" within the state of Israel, namely areas without "population." Similarly, on maps depicting the distribution of employment in Israel or "universities in Israel," only the Israelis who are employed in the occupied Palestinian territories and Jewish universities or university-extensions in the occupied territories are depicted: There is no trace of Palestinian universities such as Al-Kuds or Bir-Zeit or the University of Bethlehem, which may be much bigger and much more known, or of Palestinian employees, who are characterized in this book as "foreigners" or "host workers"—namely as illegal invaders or temporary human labor force: "Part of the *foreign workers* are Palestinians. . . . They are employed in unprofessional jobs and their wages are lower than that of the Israeli citizens who work in the same jobs. . . . *This is characteristic of all developed countries*" (Fine, Segev, and Lavi 2002, 32; emphasis added).

Treating the Palestinians as foreigners points to an odd geographical perception: The Palestinian territories are presented as part of Israel, and yet the inhabitants of these same territories are foreigners. However, the readers may not be aware of this peculiarity because the territories are not marked as Palestinian.

In order to inculcate the idea that Israel's borders are subject to "constant dynamic changes" and are nothing more than a result of an accidental ceasefire decision, *The Geography of the Land of Israel* presents a map titled "Israel after the Oslo Agreements" (17). The text on the right side of the map explains, "It was decided to establish a Palestinian Authority in *some areas* of *Judea, Samaria* and the Gaza region. Those were not drawn on the maps in this book because they are not international borders yet" (emphasis added).

Calling the West Bank by its biblical Hebrew name, Judea and Samaria, on spite of the fact that these areas were never annexed to the state of Israel, reinforces the well-rooted notion that these areas have been "ours" since time began. Students are not asked, nor do they question, why international decisions, regarding the annexing of the Golan Heights and the demarcation of Palestinian-controlled areas are not respected.

Mind Infection

Israeli children are deprived of the knowledge of the geopolitical facts of their country and of their immediate neighbors, their history, their culture, and their merits. Israeli children are educated to see their neighbors as an unwanted element. For me this is not education, this is what the scientist Richard Dawkins (2003) calls "mind infection."

Dawkins speaks about viruses that infect the mind. Children, he says, because their minds are gullible and open to almost any suggestion, are not immune to mental infections of all sorts of propaganda and fashion. They are easily persuaded to pierce their faces and tattoo their bottoms, to turn their hats around and bare their bellies, and to believe in angels and fairies. They are equally open to acquire political beliefs and to appropriate mental maps that will later influence their decisions on the question of the future borders of the state.

Israeli children are mind-infected from an early age. By the time they are old enough to become *real* soldiers, they have already learned to be *good* soldiers, which means their minds are totally infected and they are incapable of questioning the "truth" that has been inculcated in them. This is part of the explanation one can give to the terrible deeds that are committed every day by good Israeli boys and girls, who are characterized once and again in the popular press, like the newspaper Haaretz, as "people of values." Therefore, it is high time to ask, what values do they uphold?

The following lines are part of a personal preface to a term paper by Tal Sela, analyzing a history textbook. Tal Sela is a member of the Israeli-Palestinian organization Combatants for Peace and one of my university students:

On the 5th of September 1997 I found myself in Lebanon, on a rescuing mission. All my friends were in the battle, twelve soldiers were killed. The following days I was happy: "I am alive, I survived," I said to myself. However, a year later I was in deep depression. Sad and morose, I decided to consult a psychologist. After a few sessions I was able to gather up my forces again, both physical and moral. I could reorganize my thoughts. Then I understood that the mental crisis I had was in fact a moral crisis, a crisis of consciousness. What I actually felt was frustration, shame and anger. . . .

How could I be so gullible and let myself be duped? How can I explain that a man of peace, a philanthropist, exposes himself to such a morbid experience of his own free will? Today, like every two weeks, I drove peace activists to the military check-points of the Israeli army in the occupied Palestinian territories. I saw an officer put tight handcuffs on both wrists of a taxi driver because he failed to obey the soldiers' order to park here and not there. "We told him a thousand times," the soldiers said. The man was lying on the ground in the worst heat of the summer, thirsty, for hours on end. His friend was luckier: he had to stand on his feet, in a cell, without handcuffs.

What pushes these young Israeli boys to play the role of supreme judges until they lose all judgment? In my opinion it is the Grand Zionist Narrative, which serves as a collective conscience to the whole Israeli society, explicitly as well as implicitly. This grand narrative is the system of values that makes us belong to this particular collective. This is the system that dictates the relationships between us and the Palestinians. This is the social structure that dictates the relationships between any grand narratives: Hutu and Tutsi, Greek and Turks, Sephardic Jews and Ashkenazi Jews, Catholics and Protestants, blacks and whites, men and women.

How else can one explain young people who were educated to love their neighbor as they love themselves killing their neighbors, torturing their children and their elders, uprooting their olive trees and their vines, demolishing their houses, poisoning their wells, destroying their educational institutions, their libraries, and their hospitals, for no apparent reason other than their being neighbors? The only explanation is that their minds are infected by parents, teachers, and leaders who convince them that the others are not as human as "we" are and therefore killing them is not real killing; it has other legitimating names such as "cleansing," "purifying," "punishment," "operation," "mission," "campaign," and "war."

Although I speak about Israeli boys, this is not an Israeli affair because the epidemic is worldwide. My nephew, Doroni, 7 years old, who lives in the United States, came home on Halloween day and said he wanted to be a soldier and then to go to "Eye-raq" and save America. How many American young men, ignorant as he is of the absurdity of this statement, actually went to Iraq and died there without knowing why, but with the words "save America" on their lips? The question is, how were these false values imprinted on their minds and how can they be erased?

The human psyche, says Dawkins, has two great sicknesses: the urge to carry vendetta across generations and the tendency to fasten group labels on people rather than see them as individuals. We are all suffering to death because of labels, but only a few of us know that the way to fight labels is to refuse labels. The way to defeat false value systems is to expose them.

The viruses of the mind are only partially weakened by young people like Tal Sela and other Israeli 'refusniks' (soldiers who refuse to serve in the army or in missions that are immoral) such as in the group Combatants for Peace. But how can we free mind infected children from the grip of those viruses before they find their final rest in the ever-growing, underground kingdom of dead children? How can we educate them that it doesn't matter whether their head was bare or not in a synagogue or a church or a mosque, whether they were circumcised or not, whether they pronounced forbidden words, or ate pig or cow, or whether they had a hot milky chocolate after their salami pizza just before they were blown up by someone who didn't?

The children of today are the parents, the teachers, the politicians, and the generals of tomorrow, just as the generals and politicians and the mothers and fathers of today were the schoolchildren of yesterday. In countries where patriotism is a supreme value mothers raise their children with all love and care in order to sacrifice them to the god of death, and fathers urge their children to commit themselves to armies whose orders include words and actions such as "selection," "forced expulsion," "cleansing" or "purifying" regarding human beings.

Change should come from conscientious adults, teachers and parents. It is time to teach our children throughout the world to be skeptical, to mistrust words of "value," words such as "loyalty," which actually mean collaboration in crimes against humanity, words such as "democracy," "security," and "safety," which are used as an excuse for murder, for robbery, and for the construction of ghettos where people are doomed to die for lack of medical services, education, or livelihood.

It is a terribly hard task for people who were educated in Israel or in the United States or in any other "Western" "democratic" country to admit we were raised on false values and to teach our children to disrespect authority. The only thing that can enhance such a change of spirit is the constant image of the mutilated small bodies of our dead children. Only by looking at them can we realize we must ask their forgiveness for not refusing the evil viruses before and for letting them be the victims of the horrible mental infection we are all suffering from—to look at their innocent, astonished, disillusioned small faces and ask ourselves, after the poetess Anna Akhmatova, who also lost her son to an evil inhuman regime, "Why does that streak of blood rip the petal of their cheek?"

References

Aharony., Y., and T. Sagi. 2002. *The geography of the land of Israel—A geography textbook for grades 11–12.* Tel-Aviv. Lilach.

Bar-Gal, Y. 1993a. *Homeland and geography in a hundred years of Zionist education.* Tel-Aviv: Am Oved.

———. 1993b. Boundaries as a topic in geographic education: The case of Israel. *Political Geography* 12 (5): 421–35.

———. 1996. Ideological propaganda in maps and geographical education. In *Innovation in geographical education,* ed. J. van der Schee and H. Trimp, 67–79. The Hague: Netherlands Geographical Studies, IGU, Commission on Geographical Education.

———. 2000. Values and ideologies in place descriptions. *Erdkunde: Archive for scientific geography* 54(2).

———. 2003. Geographic politics and geographic education. Speech given to the Conference of the Geographic Society, Bar-Illan University, Tel-Aviv, Israel.

Bar-Navi, E. 1998. *The 20th century—A History of the people of Israel in the last generations, for grades 10-12.* Tel-Aviv. Mapa.

Bar-Navi, E., and E. Nave. 1999. *Modern times part II—The history of the people of Israel. For grades 10–12.* Tel-Aviv. Mapa.

Dawking, Richard. 2003. *A Devil's Chaplain. Selected Essays.* U.K.: Weidenfeld and Nickolson.

Fine, Tz., M. Segev, and R. Lavi.2002. *Geography of the land of Israel*Tel-Aviv: Centre for Educational Technologies.

Firer, R., and S. Adwan. 2004. *The Israeli-Palestinian conflict in Israeli history and civics textbooks of both nations.* Hannover: Georg-Eckert-Institute fur internationale Schulbuchforschung.

Henrikson, A. K. 1994. The power and politics of maps. In *Reordering the World: Geopolitical perspective on the 21st century*, ed. G. J. Demko and W. B. Wood, 50–70. San Francisco: Westview Press.

Rap, E., and Tz. Fine. (1994/1998). *People in space. A Geography textbook for 9th grade* Tel-Aviv: Centre for Educational Technologies.

Notes

1. Mitzpim are Jewish top-sites: community settlements on top of mountains in the Galilee, overlooking Israeli-Arab villages.

2. Head of the Department of Geography and Environmental Studies at the University of Haifa.

PART VI

Going on the Offensive

CHAPTER 22

Interview with Thulas Nxesi, President of Education International

Mary Compton

MC: How did you become involved with the teacher trade union movement?

As a young teacher [in South Africa] in the 1980s, my dream was to build a nonracial teacher union. At that time, teachers had no right to even join a union—they had no labor rights at all. Teachers were divided along ethnic, racial, and even regional lines. These were the times at the climax of apartheid, some of it under a state of emergency where teachers were expelled from the profession for joining a union. It was worse for those who were seen to be leaders of the movement, and I lost my job. After that I joined the university and we used the university as our base for the trade union—I ended up being the general secretary of the National Education Union of South Africa, which was one of the founding organizations of SADTU.

MC: What were the main aims of SADTU at that time?

To fight for the rights of teachers to organize freely and to strike, and most important of all was the challenge of removing apartheid education policies, and in order to do that we had to forge alliances with the broader anti-apartheid movement.

MC: What was SADTU's relationship to the ANC?

We had a very close relationship with the ANC in the struggle against apartheid, but the matter is under serious debate today. Do we still need this relationship after ten years of democracy? Nevertheless, we are proud to have been part of the forces which brought about a progressive government. We expected a lot from the government in terms of workers' rights, but particularly in terms of a progressive post-apartheid education system. I think it's safe to say that we have the right policies in terms of the contents of education, but the challenge is, how do we translate these policies into practice? If you talk to teachers in South Africa, they will tell you that as much as we have had policy changes, in terms of real transformation we are still talking about overcrowded classrooms, lack of basic resources, and poor preparation of teachers. In terms of the new outcomes-based education there are heavy demands, which imply smaller class sizes and properly stocked libraries, for example—so you need huge increases in funding. But in our schools it is still a dream to talk about libraries, and if there are any, they are empty and needless to say there are hardly any computers.

MC: So why does this problem arise?

In the liberation struggles we envisaged a free, quality education for all South Africans, but we see our government backtracking on that. In fact, in the mid-1990s the government adopted a neoliberal economic strategy—which they call economic fundamentalism. This implies cutbacks in public expenditure, downsizing government, and an emphasis on private-sector imported models in the public sector, which amongst other things are imposing performance controls without the necessary tools and support to improve it. And some of these models are not appropriate to the public sector. The most obvious reason for this is that private companies are accountable to the shareholders, who are only interested in profits, and when you just try and link these models to the public service, you are dealing with two different philosophies. You are talking about improving service delivery to the people, be it health or education; therefore you cannot talk about cost cutting. Yes, you have to do it in a proper way without wasting limited resources, but cost cutting cannot be your philosophy—inherent in education is that it's labor intensive.

MC: What do you consider to be the role of Education International in tackling the neoliberal policies promoted, for example, by the World Bank?

First and foremost, I believe that Education International (EI) is its affiliates—if you don't have strong affiliates, you won't have a strong international because it would simply be an office coordinating activities, but those activities have to come from the ground. Having said that, I believe that one of EI's central functions is as an information exchange and particularly to promote research, which can be used by national unions to talk to their governments—one thing you have to accept is that the research capacity of governments is greater than ours—therefore sharing and promoting research is vital. Secondly, that research helps us to develop policies and enables us to compare and test how they have been applied in different countries and what have been the results. Thirdly, EI helps developing models for collective bargaining because in many developing countries teachers are denied the right to collective bargaining, so EI can be of help here in developing organizations like the ILO (International Labor Organization) and UNESCO in exposing some of the violators of human rights. And I think that the faces of the top officers of EI are very important for teachers in oppressed conditions—it is important that we are seen to be a voice for teachers.

MC: Does EI also have a role in offering solidarity to affiliate unions whose members are suffering?

Solidarity can take different forms—so-called "capacity building," for example, where unions from the North can help more impoverished unions to build their organizations. However in this regard it is important to emphasize one thing—international solidarity cannot be just one way—colleagues from the North have a lot to learn from the developing world and their ways of doing things. There has been a tendency in some unions of the North to follow their governments' approach to development cooperation, where it is seen more as a mechanism of social control than as real reciprocity. In fact, in my short experience of international relations, development aid is more driven by what people see as their own foreign policy interests than a real agenda to develop. But coming back to what you were saying, we have regions where there is conflict, for example, the Middle East—what is the role of EI there? It is very simple for me as a teacher from South Africa, given our historical links in terms of struggle with the Palestinians. We are very dedicated to the cause of the Palestinians. But how do we use the South African experience to help in diffusing that conflict in the Middle East? Where I am seated as president of EI, how do I interact with Israelis who are members of EI and with Palestinians? This is very crucial. We have said at EI that we will continue to oppose all forms of terrorism—including state terrorism—and that is my emphasis. Organized violence is not a preferable way of resolving conflicts, and our argument is that we need to push for a negotiated settlement in the multilateral institutions. Because conflict destabilizes all societies—it displaces children and teachers, and therefore we must take an interest in all conflicts. As educators we want peace. We must teach the new values to the young ones of tolerance of different views and different religions. We see xenophobia everywhere. The worst of it is when you see the intolerance mainly from some of the Western democracies toward what is happening in the Middle East. Worst of all are the Christians who paint everyone of the Muslim religion with the same brush as if all Muslims are terrorists. What we need to do is to focus the world on addressing the right question: what are the root causes of terrorism? Terrorism triumphs when there is poverty, where people feel they have been robbed of their birthright to live in their own land and that there's no hope for a better life

MC: Getting back to the neoliberal project that is being pursued so effectively by the World Bank, for example in South Africa. As you said, they have in many ways got all the cards in their hands: their research is further advanced, they have the media, and so on. All we have is millions of teachers all over the world trying to do a good job for the children they teach. How can EI mobilize those people?

The problem is that the project of neoliberalism does not find expression in those big offices at the World Bank—it finds its expression concretely on the ground in different countries. I see the World Trade Organization (WTO) as taking the lead in pursuing neoliberal policies in their drive to liberalize services under the GATS (General Agreement for Trade in Services). Of course it's important to raise consciousness at an international level, but the challenge is going to be with national leaders—how do they harness that energy, that support on the ground, in order to focus their organizations? I am raising this in my department at the moment in respect of the Global Campaign for Education—we must not think we will be able to do this alone. We have to build a massive coalition with the whole civil society, including the labor movement, in order to challenge the privatization of key public services—just like we did in South Africa. We were able to defeat apartheid as an anti-apartheid movement—not as different sectors. So I think the challenge to the teachers' movement is to realize that we can't continue to work alone, separate from the labor movement as a whole. And we must challenge those powerful teacher unions who are still refusing to be part of their labor centers. We must intensify that labor movement, that big coalition—how can they win their campaigns without this? To me, therefore, the onslaught of privatization of public services formed by the neoliberal agenda needs a collective response, rather than an individual response in our own little corners; and therefore, building strong organizations of workers' movements is going to help us but also linking with civil society.

MC: What exactly do you mean by civil society?

We must be very careful about so-called civil society, because these days anything and everything outside governments can call itself civil society, but they can have very different agendas. So we need to ask which of these forces are seen to be pro the working class and in defense of public services, because it is an ideological fact that there are those who want to ensure that those services are just used for the profit motive, and there are those who see them as services which are the right of all people. They must be made available because otherwise if they're commercialized, they won't be accessible to the poor, as is already happening in many countries. That's how the project is being approached in South Africa.

MC: So EI is not just about lobbying and research but also about encouraging affiliates to mobilize?

We must engage with the labor movement, but we also have to engage with the multinational institutions. We have learned in South Africa that it's not helpful to simply say, "We reject neoliberal policy, therefore we won't engage with you," because the bosses will just carry on with their program. We have to engage in the boardroom and outside—this is very important. To engage only in the boardroom is dangerous, however; you must also engage in the street.

MC: In other words, negotiate but also have a base in the street?

Exactly. For example, in the WTO, if we hadn't been there as an international trade union movement, influencing and backing our ministers and giving them the technical information they needed, they wouldn't have been able to make some of the breakthroughs they have. I mean, here in South Africa we have been the toast of civil society. It's better for people to meet with the World Bank and say, "Look, this is what is happening," than just not to talk. My argument is that engagement does not mean sellout. You are able to persuade, and even if you don't manage to persuade, at least the person knows what you are thinking because we are able to say what the alternative is. We need to move beyond just protest. We need to say there is another worldview, and we must be able to put it on the table as a counter to what they are putting.

MC: What do you see as the role of the social movements that are growing all over the world, like the World Social Forum?

It is a positive development, but we need to go beyond slogans. That remains the challenge so that we can put counterproposals to World Bank policies in the different countries.

MC: Do you see the way forward in the end without a struggle? They're not going to give up without a fight, are they?

That's why you can't rely on negotiation only. You will have to back negotiation with mass mobilization on the ground, but you can't just say you're going to mobilize, because people will ask what the counterproposal is. The two strategies work hand in hand; they enhance one another.

CHAPTER 23

The Context of Teachers' Democratic Movements in Mexico

Rodolfo Rincones

Our lives begin to end the day we become silent about things that matter.

Martin Luther King Jr.

Since July 2006 and the highly contested presidential election, individuals from all sectors of Mexican society—nongovernmental organizations, miners, farmers, students, and teachers—have formed alliances to demonstrate their discontent against the Congress and the president for enacting neoliberal reforms. Government approval to reform the social security programs ignited massive protests among rank-and-file teachers in several states, even in states where teacher protests were traditionally minimal or absent. Wage-earning Mexicans have been devastated by neoliberal reforms. Since the late 1970s the minimum wage has fallen by 75 percent. During the 2000–2006 presidential period alone, the minimum wage fell by 22 percent. Ten million workers, 24 percent of the economically active population, make the minimum wage or less. Fifty million Mexicans live below the poverty line (Roman and Velasco 2007).

Historically in Mexico there has been fear of democratic and independent unionism. The PRI (*Partido Revolucionario Institucional*) ruled Mexico for over seventy years, enacting labor laws that gave the state control over union recognition and the right to strike. The PRI integrated into itself the only officially recognized unions, exercised authoritarian control, and repressed union members who disagreed with or protested against autocratic union leaders. This allowed gains for limited sectors of the working class—and rampant corruption by union leaders.

Labor laws and employment practices have undermined the creation of independent unions, most notably empowering employers to designate the workers with whom they wish to negotiate. Employers thus negotiate with fake or false unions, allowing union officials, known as *charros*, to sign collective agreements whose contents are unknown to the workers. Of the approximately 42 million workers in Mexico, probably only 13 percent are unionized, of which 90 percent belong to fake or false unions (Alcalde 2007).

In 1943 the government created SNTE (Sindicato Nacional de Trabajadores de la Educación), the national teacher union, with three aims: to eliminate the influence of other teachers' organizations that had been struggling to form democratic and independent unions since the 1920s; to subdue the influence of the Catholic Church; and to restrain the participation of parents and entrepreneurs in education.

SNTE's economic power is immense. It receives one percent of the salaries of all the members (Vergara 2007) in addition to discretionary funds transferred from the federal government. Because of its authoritarian structure and government protection, SNTE leaders are not accountable for the management of these funds. Pablo Latapí (2004), one of the most notable Mexican educators, notes that SNTE exercises absolute professional and political control over its members. Teachers depend on SNTE to receive a position, transfers, salaries, benefits, credits, retirement benefits, scholarships, and home mortgages. For this reason, Latapí refers to teaching as the subjugated profession. Most new teaching openings are controlled by SNTE, and newly appointed teachers are subordinated to the union's power, since teaching appointments are made for life and can be inherited and even sold.

Further, SNTE oversees and controls *normales* (normal schools, teacher training institutes) and has used its political power to prevent universities from offering programs of teacher preparation. Teachers who graduate from normal schools have first access to new teaching positions, and it is commonly accepted that only loyal union members get appointed to principal and supervisory positions. In addition, SNTE is seen as having little moral credibility because of its defense of members who have embezzled school funds, committed fraud, and committed crimes such as sexual abuse of students or teachers. Finally, SNTE has consistently opposed evaluations of teaching and social and community participation in schools, alleging that parents are an obstacle in the educational process.

In 1979, CNTE, (*Coordinadora Nacional de los Trabajadores de la Educación*) was constituted and emerged as the most prominent dissident movement. CNTE is an organized national alliance of dissident teachers in reform caucuses and official union sections within SNTE. CNTE is very militant and often has deep community roots and engagement in broad, popular struggles, as was the case in Oaxaca in 2006. Yet it appears that support from society may have diminished, and while the demand for union democratization continues, it too has lost ground. Fuentes Molinar argues that CNTE is fracturing because it tends to reproduce several of the practices from the SNTE leadership (Herrera Beltrán 2005). He suggests CNTE needs to revise its role and strategies to regain the leadership it once had in the teacher dissident movement, to be able to defend the assault on public education.

Section 22 of SNTE in the state of Oaxaca has carved out great autonomy in decades of struggle against the national leadership and has played the leading role in the Oaxaca movement. Section 22, along with a variety of popular organizations and CNTE, formed a coalition that carried out a militant urban resistance against neoliberal reforms—until it was repressed on November 25, 2006, by state and federal police and armed forces.

Besides CNTE several other smaller but nonetheless important dissident groups have emerged: SINPPE (*Sindicato Independiente Nacional de Profesores y Personal Educativo*), created May 17, 2007, as an alternative to SNTE in fourteen states; STIPN (*Sindicato de Trabajadores del Instituto Politécnico Nacional*); SITET (*Sindicato Independiente de Trabajadores de la Educación de Tabasco*); SITECIM (*Sindicato Independiente de Trabajadores de la Educación de la Ciudad de México*); MDMZ (*Movimiento Democrático Magisterial de Zacatecas*), dissident group against Section 34 of SNTE; *Fracción Democrática en Aguascalientes*; *Acción Ciudadana del Magisterio en Baja California*, and *Magisterio Democrático del Estado de México*.

Looking Forward

President Felipe Calderón has initiated fierce repression against social movements and protests, and the Secretary of the Interior is known for the tough handling of the anti–corporate globalization protests in Guadalajara on May 2004. The current government is marked by a tendency to criminalize protests and other types of demonstrations, including recently those of miners, flight attendants, and public employees. Yet the new government faces a lack of legitimacy due to the popular viewpoint that its victory was a result of massive fraud, in addition to widespread anger at decades of market-based reforms that have impacted negatively living standards and lack of control of the new Congress.

I suggest that in this context the teachers' movement has to develop strategies of continent-wide solidarity, strengthening existing alliances and coalitions. In my opinion, instead of fighting only to take over corrupt SNTE sections, it is also necessary to develop a new political culture where teachers develop and deepen their understanding of democracy and reflect on the potential and limits of unionism. The most pressing need, however, is for teachers to reflect upon their practice and role in transforming the lives of students and parents in all corners of the country.

References

Alcalde Justiniano, A. 2007. Contrapunto. Televisa, August 16.

Herrera Beltrán, C. 2005. Alarmante injerencia del SNTE en la educación pública: Fuentes Molinar. La Jornada, (Mexico, DF), May 15.

Latapí, P. 2004. *La SEP por dentro: Las políticas de la Secretaría de Educación en Pública comentadas por cuatro secretarios (1992–2004)*. México: FCE.

Roman, R., and E. Velasco. 2007. Mexican workers call for a continental workers' campaign for living wages and social justice. Global Research, May 12. http://www.globalresearch.ca/index.php?context=va&aid=5625 (accessed August 21 2007).

Vergara, R. 2007. Todo el control, todo el dinero. *Proceso* 605 (August 5): 12–16.

CHAPTER 24

In Mexico, to Defend Education as a Social Right, We Must Fight for Union Democracy

María de la Luz Arriaga Lemus

In Mexico we fight for democracy in our labor unions as part of our work to defend education as a social right.

María de la Luz Arriaga Lemus

As did many young normalistas (teachers) from the 1968 student movement, when I completed my studies at the age of seventeen, I entered the teaching profession (in a fifth-grade elementary school classroom) not only determined to be an excellent teacher but also with a clear intention to work toward democratic and social transformation. In 1968 one of the most important student movements in Mexico's contemporary history emerged, clearly aimed at defending democratic freedoms. While the 1968 student movement was a determining factor in my later decision to join the fight for democracy in Mexico, my normalista teacher training was also an important influence, since it was characterized by instilling in us a social commitment to our students and to our country. It is important to remember that in Mexico education is a social right, won in the revolutionary struggle of 1910. But in addition, teachers have historically been social leaders in our communities.

I was fortunate to work in a marvelous part of Mexico City called Cuajimalpa, which at that time was still a small town. I say I was fortunate because of a few circumstances in Cuajimalpa: there were many other young teachers like me, and it was a community with a clear social structure of solidarity in which teachers played an important role, and it was a community with many unsatisfied needs.

With great creativity and enjoyment in teaching, we began a series of initiatives in classrooms, in schools, and in the community—and our actions quickly drew the attention of authorities in the schools and in the school district. And to my great surprise, these authorities tried to prevent us from continuing such initiatives. Here are some examples:

To complement school activities, we acquired a piece of land where the children could go every week to plant in the field, and then we would go to a nearby river. But school authorities began to limit activities outside the school building. We began a literacy campaign in the community and we used the school building—which seemed completely normal to us and one of the purposes of a school building. However, the school director prevented us from using the school, and we had to instead use public parks and the church grounds.

In addition, educational authorities began to harass us. For those of us who worked the hardest, who worked in literacy campaigns on Saturdays, who used our extra time for extracurricular activities without any type of extra pay, who collectively questioned the authoritarian educational model that was imposed on us, and who sought new alternatives—extra bureaucratic demands were placed on us, and there was greater control over when we entered and left the school buildings. The intention of school authorities was to control us and limit us to a more traditional role of a teacher who arrives at school, gives classes, and returns home—without any real commitment to students.

But above all, they wanted us to accept the marginalized role imposed on us. We had to comply with the imposed plans and programs and with authoritarian teaching methods. If we demonstrated an interest or demanded participation in the decisions of what to teach and how to teach, we were seen as presenting a danger—since this was only one step away from demanding the democratization of schools, of the teacher's labor union, and of the country.

And naturally, if our work as teachers made them afraid, they were terrorized when we managed to create close links with parents. The entire educational system in Mexico is exclusive in nature: official discourse claims that parents should participate in the educational system, but parents are only asked to make economic contributions and are asked to contribute little or nothing to decisions regarding the education of their children.

Hundreds of thousands of teachers in Mexico find themselves between two fires: the harassment by educational authorities in their bureaucratic, authoritarian administration and the relentless pursuit by the official or *charro* labor union leaders, who prevent teachers from organizing and collectively participating in defending our rights and defending public education.

SNTE and Teachers' Ideals

In order to defend the right to education in Mexico, it has been indispensable to fight for democracy in our labor unions. At the moment we begin to work as teachers, we are automatically affiliated with SNTE, the National Educational Workers Union (*Sindicato Nacional de Trabajadores de la Educación*), and the leaders of this labor union at the national level and in the locals become an obstacle when teachers want to improve the quality of education and improve their working conditions. Up until the second half of the 1980s, education for elementary and preschool teachers consisted of three-year programs following junior high (*secundaria*) school. This training was provided in urban and rural *normales* (teacher training schools). When we finished our training, at eighteen to twenty years of age, we were assured of employment as teachers in public education, and the federal government assigned each of us to a school.

Mexican teachers were for many years fundamentally the daughters and sons of rural, working class and low middle class families. After the democratic teachers' movement arose and grouped together in CNTE, one way of changing the social composition of Mexican teachers as a group was to make finishing senior high (*bachillerato*) school obligatory before entering *normalista* teachers' training, thus increasing the education required for teachers by four years. Another way was to begin to close the *normales*, primarily those in rural areas.

The official labor unions in our country are part of the political system, and they are an instrument of the state used to exercise its power. As a result, any fight to stand up for our rights or to demand our right to elect our representatives must confront the state apparatus.

Currently, official labor union leaders, particularly those in SNTE, are those promoting all the neoliberal policies that have been imposed on us in the educational sector. I am speaking here of policies to keep wages low and to keep employment precarious, to limit budgets, and to decentralize education as a condition for the sector's privatization. And I am speaking of the segmentation of the educational system into schools of first, second and third class, with schemes in which programs for paying educational workers by their "merit" play an important role, as do the so-called "quality schools." We are also suffering from an avalanche of standardized tests as instruments for excluding students, as well as changes being made in curricula, in order to suppress critical education with social commitment, imposing instead the commercial values of individualism.

SNTE is the largest labor union in Latin America. Its totally authoritarian, antidemocratic structure is controlled by a powerful group with strong influence on national politics, a leadership that violates its own bylaws to the point of illegitimately naming the current president, Elba Esther Gordillo, as the union's president for life. The power enjoyed by the political group that controls SNTE was demonstrated in the July 2006 election process. SNTE was not only able to place a significant number of its representatives in positions in the House of Representatives and the Senate, but it was involved in the electoral fraud that brought Felipe Calderón Hinojosa, the candidate of the right wing, to the presidency. This same powerful group was then able to obtain political and financial positions in the new presidential cabinet, one in the Ministry of Public Education, another in the Social Security Institute for State Workers, and another in the National Lottery.

For seventy years SNTE provided the social base for the PRI, the Institutional Revolutionary Party (*Partido Revolucionario Institucional*), the reigning political party. But from 2000 to 2006, Elba Esther Gordillo and her group became the primary support within the labor unions for right-wing President Vicente Fox. Her group was expelled from the PRI, formed a new political party, and as an ally to the right-wing PAN party became a key player in the electoral fraud.

Thousands of teachers have come together to form CNTE, the National Coordination of Educational Workers (*Coordinadora Nacional de Trabajadores de la Educación*). CNTE has struggled for twenty-eight years to democratize the union and defend public education as a social right—specifically, to improve the conditions for children to study, to expand access to education to all Mexicans from preschool age to higher education, to defend the right to education without cost, and to avoid the privatization of education. It has also worked to build an alternative model for education under high-level democratic, scientific principles, in the service of the Mexican people. These actions, organized among broad sectors of teachers, are an expression of grassroots democracy. Through strikes, picketing, protest marches, occupations of highways and radio and television stations, and a multitude of mobilizations and creative expressions of struggle, we have been able to prevent neoliberal reforms of the educational sector from advancing more rapidly.

And of course, in all these years, democratic teachers have suffered constant repression: from administrative sanctions to jail and murder. The most recent example is the repression in the state of Oaxaca, where a struggle initiated by Local 22 of SNTE (with 60,000 teachers) for an increase in salaries and better conditions for students was violently repressed. This gave way to the emergence of a movement of teachers and the general population that has been exemplary as an exercise in grassroots democracy, with democratic forms of organizing and taking actions. It is also an example of terrible repression that has caused deaths and hundreds of injured and imprisoned among participating sectors: teachers, representatives of indigenous communities, state workers, parents, journalists, students, and others (Arriaga Lemus 2002).

What We've Learned in the Struggle

The first lesson we have learned from the struggle has been that democracy is a necessary step in the struggle for education as a social right. This is a historic struggle that cannot be resolved with only a single demand being satisfied, and this is a struggle that extends beyond a single sector. We have also learned that creating links with parents and with students is vital in defending this fundamental human and social right. We have also learned that unity among the various contingents of educational workers—from those working at the preschool level to those in universities—is necessary if we want to move forward in achieving our demands.

We have learned in this struggle to recognize our friends and our enemies, and we have also constructed paths of solidarity and exchanges that have transcended borders.

Free trade agreements imposed on us by governments and transnational corporations as part of their neoliberal globalization agenda have led us to create the Tri-National Coalition in Defense of Public Education and the Civil Society Network for Public Education in the Americas, as Larry Kuehn describes in his essay. Participating in the Mexican section of the Tri-National Coalition are international unionists and representatives of labor unions and union locals in the educational system, encompassing basic to higher education, as well as social and student organizations from the entire continent.

This process has allowed Mexico's democratic teachers to become more visible, since SNTE participates in international organizations of education unions, and only the national executive committee can send its representatives. It was previously essentially impossible to bring the antidemocratic situation and the repression in Mexico to the attention of labor unions in other countries. But formation of the Coalition and the Network[1], strengthened our struggle in defending public education and building democratic societies through solidarity and joint actions. They have served as powerful instruments in improving understanding among educational workers in the three countries and in the Americas, in expanding knowledge of our different realities, in developing documents and alternatives, in conducting simultaneous campaigns to bring an end to the repressive actions of governments, in expressing our disagreement with the social policies of international entities (especially educational policies), and also in bringing visibility to our alternative proposals.

In Mexico, as in other countries around the world, globalizing neoliberal policies are increasingly deteriorating and impoverishing public education systems. In Mexico, the organization that should be fighting the imposition of these policies has instead been supporting and promoting them, and I am speaking here of SNTE. But I would also like to highlight the combative spirit and determination on the part of thousands of teachers at all levels of education, as well as educational workers in general and parents and students who have engaged in heroic struggles and have developed countless ways to organize and defend education as a social right.

Reference

Arriaga Lemus, M. 2002. Impacto Político de las luchas magisteriales en México, Masters thesis, UNAM, México.

Note

1. More information can be found on the websites of the Mexican section of the Tri-National Coalition (www.forolaboral.com.mx) and of the Red-SEPA (http://www.red-sepa.net/ideas/).

CHAPTER 25

A History of the Search for Teacher Unity in South Africa

Harold Samuel

Stories of South Africa cannot be told separately from the struggle for freedom. The search for teacher unity is no exception. Indeed it could be seen as central to the struggle because the organized teaching profession was seen by many as an ideal platform from which the state could be challenged.

In the half century following the Union of South Africa in 1910, the country searched for its place in the sun—developing an administration to maximize the use of its rich natural resources and joining the allied nations fighting the two world wars against an aggressor striving to dominate Europe.

There are many records of the heroism of white South Africans in these wars, but little is said of the people of color who also joined the army, including two of my own uncles, Douglas Isaacs and Sylvester Henriques. Although their fight helped to bring about peace in Europe in 1945, when they returned home, they were faced with a weird policy of segregation and domination: apartheid.

The 1948 general elections saw the Nationalist Party sweep into power to inflict its warped ideology focused on the advancement of the ruling class with the rest being subjugated, separated, and dominated. But through the sweat and toil of the people it had chosen to label non-European or non-White, South Africa progressed to emerge as a powerful partner popular with countries in the Western world.

The state's master plan was to dominate through a well-orchestrated school system: the Bantu Education Act of 1953 saw education carved up into racial blocks. Educators were cleverly divided into nineteen separate departments of education, and control was carefully maintained through an effective bureaucracy. Not surprisingly, the way in which education was offered to the South African people was woven through the fabric of the political struggle against apartheid.

The nation had moved onto a slippery slide, and despite all its grand plans for separate development, the disenfranchised strove for liberation. The struggle for freedom is well documented in many writings, and this chapter will not travel down that road. Rather, this chapter attempts to trace the organizational and political divisions that dogged the search for unity among all educators in South Africa and their realization that the future lay in one teacher body for all of the country's educators.

Movement toward Unity

Concerned with parity of salaries and conditions of service, Indian and Colored teachers ("Colored" was the term used by the apartheid regime to denote people of mixed race) identified the need to form a united teaching force. On November 1, 1958, they established a federation of seven different teacher associations, called the South African Federation of Teacher Associations (SAFTA). But it took another sixteen years before they were able to convince any White teachers to join their ranks. And even with the addition of three progressive White teacher unions in 1974, the federation was doomed to failure without the inclusion of African teachers. The powerful African Teachers Association of South Africa (ATASA) had already obtained recognition and affiliation to a world body of teacher unions, the World Confederation of the Organization of the Teaching Profession (WCOTP), and was unwilling to join. Black consciousness appealed to them, and they identified SAFTA as only a convenient talking shop. Since the federation failed to embrace all within a unitary organization, no one was surprised when it disbanded in 1984.

Early Days in the Struggle

I remember as a young teacher at Stanger High School in the 1950s attending the annual general meeting of my branch at the school. It was quietly announced that Chief Albert Luthuli—a resident in a nearby "native reserve," was to be the guest speaker. I remember him speaking from handwritten sheets of paper when he encouraged the membership to excellence in their work. It was the first time I had listened to a speaker from across the racial divide, and I was very comfortable with the address, which quoted Ghandian philosophy. Inviting the chief to speak was considered very daring at the time. Much later it became fashionable among many of the fragmented teachers' collectives for an exchange of visits to their annual conferences. These gatherings became pleasant social events, with a common thread for improvements in conditions of service encouraging this collegiality. Pressure was exerted on the authorities for meaningful change on many different fronts.

It was largely accepted that with the dissolution of SAFTA a single unitary collective of educators was the only way to go. Visionaries realized that this was the first step toward the ultimate goal of a united nation for all the people of this land. The African National Congress (ANC) in exile gave the impetus for unity by encouraging teachers' leaders to meet across the color line and across the South African borders. Historic meetings planned with clever subterfuge—since passports were problematic—were held in Tanzania, Zambia, and Zimbabwe. Educators were convinced that despite the rigidity of the laws, unity among professionals was possible.

Many anxious moments were spent debating the merits of unionism vs. federalism. Many regarded unionism as being in opposition to professionalism. In South Africa, the White collectives were uncomfortable with the radicalism of unionism and favored federalism, which allowed them to retain their identities. It was inevitable that this split the teaching force in two. The White federalists took their lead from federal models in the United Kingdom and the United States, with their provincial units now making up a Teachers' Federal Council. Those who chose federalism later emerged as the National Professional Teachers' Organisation of South Africa (NAPTOSA).

The unionists, on the other hand, set up a new grouping in the 1980s, the nonracial National Education Union of South Africa (NEUSA), loosely spread across the whole country. NEUSA led the unionist thinking as a counter to the professional approach. They succeeded in attracting the attention of other smaller groupings with a common unionist ideology, eventually leading to the Harare Accord of 1988.

Harare Accord 1988

The following statement from *The Dissolution Journal* of the Teachers' Association of South Africa (TASA) challenges the theory of federalism.

Retaining our identity through federalism is entirely unacceptable in a truly democratic new order. Ethnicity, federalism and minority grouping do find favor with those who are uncertain about the future. A single unitary system of education is a certainty for the future and teachers, as leaders in community life, must set the lead in this direction. If we are truly committed to preparing for democracy and a new order, a single united teaching force is the best possible investment for the future. In spite of the threats of non-recognition by the state, we must continue with the struggle and victory will surely be ours. The struggle availeth naught if we merely wait passively for recognition.

I was the last president of TASA, the national Indian collective, and the directions to be taken were arrived at through the democratic processes of its constitution. There was a minor concern about the assets that TASA had accumulated, but its dissolution conference was clear minded when it agreed that the assets in total should be moved to the newly established South African Democratic Teachers Union (SADTU) in October 1990.

A watershed meeting took place at Harare, in Zimbabwe, in April 1988. The meeting had been convened by the Congress of South African Trade Unions (COSATU) with the support of the government in exile. It was a historic moment and expectations were high. Support had also been forthcoming from friends overseas, including sponsorship from the International Federation of Free Teachers' Unions (IFFTU). Largely through the skill and expertise of the general secretary, Comrade Jay Naidoo (General Secretary of COSATU), consensus was reached on the key issues—including ten underlying principles for teacher unity. Nine organizations signed the Harare Accord for teacher union, with two preferring observer status.

It is a credit to the leaders of TASA that the question of loss of status in a merger did not enter into the equation at all. Everyone on the TASA executive knew that there was little possibility of being drafted into the national leadership of SADTU because our numbers were much smaller. It was a selfless act, as they realized that the establishment of SADTU with all it implied for a unified teaching force as a stepping stone to a unified country, was far more important than the aspiration for positions.

The SADTU Launch

The Harare Accord led to the establishment of the National Teachers Unity Forum (NTUF). Unity talks across the borders and in South Africa closed in 1988 and culminated in the launch of SADTU in Johannesburg on October 6, 1990, a historic moment in the search for unity among the teachers of South Africa.

A celebratory rally was held at the Orlando Stadium in Soweto on the following day. A signal honor was the presence of the president of the world body WCOTP at both functions. Mary Futrell delivered sterling messages conveying support from internationals across the globe. The ecstatic mass of delegates ululated with song and dance at this significant moment. A blow against apartheid had been struck, and it was only a matter of time before challenges from other fronts would follow. The launch and rally gave educators their first taste of freedom and a sense that the portals at the Union Building in Pretoria were beginning to tremble.

The Struggle Continued

There can be no doubt that the unity of teachers under SADTU may be viewed as contrived to achieve a political goal. Consequently, the state as the third force denied SADTU official recognition and did its best to cripple the union.

I had the honor to be elected the National Treasurer, and as such I had a mighty task on my hands. There were difficulties in funding in the prelaunch period when each organization had to carry its own expenses. With the launch completed, inexorable expenses followed, and we were compelled to look for assistance from elsewhere. A call went out to friends overseas, and both WCOTP and IFFTU responded positively. While WCOTP funded the launch and rally, the latter sent us a sizable advance of U.S.$40,000 for governance thereafter.

The broad democratic movement locally and overseas lauded the launch as a precursor to a single united nation for all. This was too progressive for many in government, which was still trying to justify its apartheid existence. Great resistance followed, and their strategy was to be obstructionist wherever possible. They hid behind draconian laws and hit us where it hurt the most: they prevented us from raising funds through subscriptions (dues) from the paychecks of members, setting the stage for another collision course.

Protest meetings, rallies, marches, and demonstrations followed, and the membership sprung into action, confronting local and national departmental levels. The lifeblood of the union was being squeezed, and the cost of running the national office and fourteen regional structures was considerable. It was during this crisis that our friends from overseas became even more supportive. There were times when funding was delayed, but the rent and salaries had to be paid. We were stretched to the limits, and our greatest embarrassment came when furniture in the national office had to be repossessed! There were times when the staff was forced into delayed salary payments, while some senior staff members were prevailed upon to wait even later.

The first national strike in the history of education in South Africa was called in May 1992. The key issues were the disparity in salaries paid racially and the nonrecognition of the union. The state had a hard time countering these arguments. They could not win, and slowly but surely the struggle for recognition was granted from the nineteen differing departments of education. The granite block crumbled, and recognition followed grudgingly. The shackles were broken. Subscriptions flowed through deduction from salary, and the membership grew at a phenomenal rate. The collective political pressure exerted on Pretoria brought the results we desired.

Self-Sufficiency

The last comment on funding records the gratitude we owed to the many unseen faces overseas who came to our rescue in our darkest hours. In those trying times, we were able to raise only 12 percent of our income from local subscriptions. Had teachers in Norway, Sweden, and Canada been slow to respond to our appeals, we would surely have ground to a halt. It is difficult to picture how much of an effort would have been required again for a process of resuscitation, but we certainly would not have been in place to welcome the new democracy in 1994. Once recognition was granted, recruitment accelerated, and the finances became transformed. Local income grew to 41 percent in 1994. In 1995 only 12 percent of union income was received from overseas. Local contributions reached 88 percent from the ten Rand subscription received per month from 80,000 members.

This turnaround amazed us and confounded our international friends as well. We were reliably informed that this was unique in the annals of donor funding. Many other emergent national collectives, some over twenty-five years in existence, could not subsist on their own and were forced to depend heavily on donors. Another special plaudit had been our record of sound corporate governance, coupled with strict financial control. We had paved the way for other emergent nationals. SADTU was invited by sister organizations in Africa to share our success story with them. I enjoyed being part of the struggle that helped the organization to be where it is today.

Unionism versus Professionalism

It is a fact that in more recent times both Indian and Colored teacher membership has been on the decline. The union takes its strength from a largely

African membership, but this is not unnatural given the demographics of the country. Some Indian and Colored teachers are more comfortable being in "professional" collectives, which are less militant on labor issues. This is a paradox, since SADTU has a powerful education wing, which enjoys the largest allocation of its national expenditure. The fact that SADTU chooses to use its power through industrial action such as strikes makes many individuals uncomfortable, but the successes of the union at the bargaining table ("chamber" as it is called in South Africa) for salary increases are readily accepted by all, with little thought about their affiliation.

It is a disappointment that Indian and Colored teachers are uncomfortable with SADTU. Many of the younger members who have joined the profession do not have a true understanding of the past. A healthy opposition to the ANC, the ruling party in South Africa, also contributes to the division in affiliation loyalties. Many see the alliance of SADTU with the ANC as a weakness and choose the opposition as their best comfort. Yet this is part of the democratic process, and I am hopeful that in time a more realistic balance will arise.

And despite its differences with SADTU, NAPTOSA reaches consensus with SADTU on many issues, especially when representatives meet with the employer at the bargaining table. There is also a growing tendency among the professionals to espouse the use of the militancy of unionism in striving to reach their objectives—a particularly significant moment was when NAPTOSA affiliates joined the strike action called by SADTU as early as 1999.

Is it fair to conclude, then, that both groupings, unionists and professionals, are really much closer than they realize, especially when issues hinge on conditions of service? Is it fair to surmise that perhaps their ideological differences are really cosmetic? Can one conjecture that perhaps the broad scenario is really only a struggle for territory?

Reflections of the Fifth SADTU National Congress

At the Fifth National Congress of SADTU in Durban, September 2002, among the many issues discussed were two matters of special importance. The item of gender representation—the number of women in leadership does not reflect SADTU's 64 percent female membership—and the problem of unifying all of South Africa's teachers.

At the Fourth National Congress in 1998, the President of SADTU boldly challenged the issue of unity among all teachers in South Africa. SADTU membership had plateaued at 230,000 out of a total teaching force of 360,000. In addition to the defection of many Colored and Indian teachers, white teachers have also remained cocooned within their former organizations, looking askance at SADTU and being unwilling to participate in affirmative-action strategies.

Although they are uncomfortable with the union policy of affirmative action as a negotiation tool, they are overjoyed when efforts at the bargaining table successfully yield a pay rise for *all* educators in the country!

Given that SADTU has so much to offer, why has it not been able to attract teachers to its fold? It is ironic that the opposition to unionism makes a huge play for affiliation to a world body while the SADTU General Secretary, Comrade Thulas Nxesi, occupies the highest position in Education International as the incumbent President! Is the union perhaps not marketing itself correctly? But any concerted effort to influence educators of a "pinker" hue by using SADTU representatives to convince them may not have the desired results. The real problem is that many have not had the opportunity to listen to the SADTU message—they simply regard unionism as militant and confrontational. Perhaps this is an area in which our international friends could help us, although in many countries such as the United States and the UK, where teachers are divided into different unions, our struggle to be unified during the dark years of apartheid and the selflessness of many of our leaders might have something to teach them.

New Challenges

Time is said to be a good healer, but it can also be a good teacher. It can teach us how to respond to what is really important in the past, setting it aside from what was not important at all. The ultimate in wisdom is to live in the present, plan for the future, and really profit from the past. We are constantly moving through new values in business, in education, in government and in other walks of life.

Now is the time to let go of the past and cross into new areas of hope for the future. SADTU has done well but must be astute in ensuring that it does not remain locked to its past identities. It needs to move on to address deeper transformation issues beyond the agendas that gave it its birth. It will need to challenge the tensions, negotiations, successes, and failures of the present dialogues across different groupings. The leadership must test itself in new directions, must search for new objectives, and must be brave enough to ask awkward questions for which there may be challenging answers. Why do teachers join a union? Is there a need for a campaign to unite teachers under a new grouping? If not, why not?

Conclusion

George Bernard Shaw said, "Life is not a brief candle; it is a sort of splendid torch which I have got hold of for the moment and I want to make it burn as brightly as possible before handing it on to future generations." Surely it is no accident that the monogram of SADTU has the torch of learning in the fist of the nation. It is the centerpiece—a sort of splendid torch—that we have hold of for the moment and that we should want to make burn as brightly as possible before handing it on to future generations.

CHAPTER 26

British Teacher Unions and the Blair Government
Anatomy of an Abusive Relationship

Mary Compton

Try getting change in the public sector and the public services. I bear the scars on my back after two years in government and heaven knows what it will be like after a bit longer.

Tony Blair, 1999

Thus spoke Tony Blair to an audience of venture capitalists, two years into the first Labor government for almost twenty years. In spite of the "scars" inflicted by the trade unions, "getting change" in the public sector was necessary if the prime minister was to become the neoliberal trailblazer he wanted to be. And as we have seen in so many contributions to this book, this would require the disabling of the trade union movement. In this chapter I want to describe and analyze the way in which New Labor has attempted to carry out this objective in education through the policy of so-called social partnership. I believe there are lessons to be learned from this for the international teacher trade union movement—indeed some are already grappling with similar problems (see for example the controversy over the South African unions' alliance with the ANC) so my aim is to warn of the dangers of being entrapped in the way some UK teacher unions have been and to start now to set out an alternative way forward.

The Political Background

After starting my teaching career in the UK in the year that the Conservative party under Margaret Thatcher took power, I waited eighteen years for relief from cutbacks and neglect. My colleagues and I worked in a shabby building with oversized classes and insufficient teaching materials. The Parent Teachers Association had given up the traditional role of raising money for luxuries and was holding rummage sales to raise money for textbooks or volunteering to come in and paint the school. So when Labor finally came to power in 1997, many teachers and trade unionists were ecstatic. For all our skepticism about "New" Labor, at least Tony Blair promised that "Education! Education! Education!" was at the top of his agenda.

However, in the small print of their manifesto, the new administration had pledged to keep to the budget commitments of the Conservatives. As a result, for the first two years of the new government the British education system continued to suffer some of the lowest spending and the highest class sizes of any OECD country. In fact, as a proportion of Gross Domestic Product, the amount spent on education was less than it had been under the Conservatives.

What was not immediately apparent either to teachers or to their leaders was that this failure to fund education properly was part of a broader commitment to the neoliberal agenda for public services, which had begun to be introduced with such catastrophic results under Thatcher's administration. Thatcher had not only starved public education of cash but first pushed the education system toward privatization, toward crude accountability measures such as national testing and league tables (published comparisons of school test results), and toward the destruction of democratically elected local education authorities. But Tony Blair came to power determined to push further and harder than had ever been envisaged in even the most excitable dreams of the most right-wing conservatives.

The best illustration of this determination to carry through the neoliberal agenda is in Blair's policies on the organization and governance of schools. And nowhere does the Labor Party's departure from the policies for which it was founded and funded by the trade union movement appear more stark than on the issue of so-called comprehensive education. The biggest strides toward equality of opportunity in Britain were made through the development of this system of secondary education in the late 1960s and 1970s. Gone were the days in which the "eleven plus" exam divided children into academic sheep or vocational goats, their performance determining whether they attended the grammar school for the sheep or the secondary modern for the goats. Although the comprehensive project was never completed, in most areas children could now attend their local school, regardless of their academic ability. It had been Labor activists and politicians who had originally brought about comprehensive education, yet it was Labor that rekindled the notion of choice post 1997, reaching a climax when it became the main rallying call of New Labor in the 2005 general election

And there was nowhere that this empty rhetoric on choice was more emphatically and vacuously rammed home than in education. Tony Blair made a promise: "If the school that is on your doorstep is not sufficiently good, we are not going to leave you with the choice of either going privately or sticking with the school that is not up to standard" (Blair 2004).

It does not take much thought to see that this runs directly counter to the principles of comprehensive education, since it presupposes that some schools will not be "up to standard" and will presumably be attended by those children whose parents won't, can't, or aren't in a position to make a choice for a school further afield. Meanwhile, the articulate and well-heeled middle classes—the prime minister among them—leave the substandard school like the proverbial rats deserting the sinking ship.

Shot through all New Labor's education documents is the clear intention of making education increasingly orientated toward industry and business. The emphasis is on training children to be employable, on building "employers much more closely into the process of designing and delivering education and training" (DfES 2004, 73)The government's most recent policy document on schooling for children ages fourteen through nineteen makes it clear that for many children a broad education is neither necessary nor desirable. All they need are the so-called "key skills" of basic literacy and numeracy so that they can be slotted into the labor market. Take this together with the moves toward privatization, the development of so-called city academies described by Ian Murch elsewhere in this volume, and the retreat on comprehensive education, and you can see that public education as a means of helping all children to "the full development of the human personality and the strengthening of respect for human rights and fundamental freedoms" (United Nations 1948) is under a sustained attack.

Social Partnership

It is in this political context, that we must look at the social partnership agenda. Social partnership in industries as diverse as the car industry, insurance, and retail have meant the unions rubber-stamping changes that were detrimental to their members—in particular job losses. John Kelly, Professor of Management at the London School of Economics, has written extensively about the development of such social partnerships in the British Isles, pointing out that the main aim of the employers in these arrangements is to replace collective agreements with direct control by the employer, thus disempowering the unions and using them only to give credence to decisions that are against the interests of their members (Kelly 2004).

This is precisely the role that those teaching union leaders who have become embroiled in social partnership with the government have played. The setting up of the partnership agreement in education was preceded by a situation just before the general election in 2001, where schools were hit with serious teacher shortages: children were left untaught, and teachers were having to work overtime. The two largest teaching unions—the National Union of Teachers (NUT) and the National Association of Schoolmasters/Union of Women Teachers (NAS/UWT)—were engaged in some highly successful industrial action to protest the teacher shortages. At the unions' conferences that year, just before the May general election, the respective general secretaries persuaded the delegates to call off the action on the basis that the government had promised a review of teacher workload. The General Secretary of the NUT persuaded members that to go it alone would be to break all important teacher unity.

The Labor government was returned for a second term, having narrowly escaped a damaging industrial dispute, and duly set up the review. In the intervening period the review had been transformed into a workload *and remodeling* review. It rapidly became clear what "remodeling" meant. In order to release teachers from some of their work, nonteachers would be used to cover for them. Now, of course, there are some circumstances in which support staff, such as administrative and clerical staff, could do a great deal to help teachers. For example, if every department in a secondary school could have some dedicated clerical help, this would free teachers from a large burden of work. But although there was a nod in that direction, the central idea of the review was for classes to be taught by people with no qualifications at all. The most extreme manifestation of this trend was a document from a civil servant in the Department for Education entitled "Blue Skies Thinking," which foretold a school where the only qualified teacher would be the head, with all the "teaching" work carried out by unqualified people.

The report explicitly states that legally a school must have "a head teacher with qualified teacher status, but beyond that the position is very much deregulated. Gone are the days of every school having to have a full complement of qualified teachers." The report notes that "where a school does use support staff to 'teach,' the new regulations mean that this person must operate under a system of supervision by a qualified teacher—but that teacher might, of course, be the head" (National Union of Teachers 2003).

This document made clear that a large part of the rationale for such a situation is to save money: "The 2004 spending review will be very tight with new reforms largely needing to be funded from reworking existing budgets." On this model, teaching is seen simply as a system for delivering chunks of knowledge, heavily supported by computers, for which no particular training is required.

These proposals were, or at least should have been, anathema to all the teaching unions. The fight for an all-graduate profession had only been won in the last twenty years, and the use of unqualified staff to replace teachers had been fought inter alia with industrial action (see below). Therefore, although my union, the NUT, had been present at all the talks on the workload and remodeling review, we would not sell out the fundamental principle that children must be taught by qualified teachers and so refused to sign up to the agreement. Unfortunately, apart from the small and principled union of Welsh teachers (UCAC), all the other unions and associations involved were prepared to sign. As soon as the deal was signed by the two head teacher unions, the three other classroom teacher organizations, the support staff unions, the employers' organization, and the government, the signatories were signed up not only to agreeing to the principle but to promoting it among their members—and this was before the legislation had even been drafted. The White Paper, when it was eventually published, was as bad as might have been expected. The new legal framework for schools allows the head to use anyone to teach whom he considers to have "the skills, expertise and experience" to do the job. The guidance to the legislation states: "The Regulations . . . specify circumstances under which certain kinds of staff without qualified teacher status—usually support staff—may carry out 'specified work'" (DfES, 2003).

The Regulations specify the following work (DfES 2003, 7):

a) planning and preparing lessons and courses for pupils
b) delivering lessons to pupils
c) assessing the development, progress, and attainment of pupils
d) reporting on the development, progress, and attainment of pupils

So the regulations agreed to by the social partners effectively pave the way for anyone to teach—regardless of whether they have a teaching qualification. The *London Evening Standard*, for instance, reported that schools had been told to consider drafting in the Women's Institute to teach cookery and design lessons and using caretakers to teach football (Hayes 2004). When you put this development alongside the increasing privatization of education, typified by the City Academy program and its breakup of schoolteachers' pay and conditions, it does not take too much imagination for you to see either a cash-strapped head in the state system or a maverick governing body in a semi-privatized school replacing teachers with unqualified people in order to save money.

In other words, the social partnership agreement helped the Labor Government to make yet another giant leap toward the neoliberal ideal where teaching is deprofessionalized, teachers' pay and conditions put at risk, and their unions undermined. This reflects the policies promoted by the World Bank in developing countries (see examples in chapter 2). The IMF imposes a limit on the public sector wage bill and then supports a policy of paraprofessional teachers who have been trained for a few weeks, put into schools, and paid a third of teachers' salaries.

It is difficult to understand why the other teaching unions were seduced into signing the agreement. Of course, there were gains for teachers. The guarantee of 10 percent so-called Planning, Preparation and Assessment (PPA) time for all teachers was a big step forward, particularly for those working in primary schools. However, as I have shown, this was bought at a price that many teachers considered to be much too high. And had the unions remained united on this issue, it would have been possible to avoid selling out this fundamental principle; indeed had they been prepared to reactivate their joint industrial action and enlist the support of parents it is highly likely that the government's agenda would have been rolled back. However, the other unions were willing to gamble on the inertia of their own membership (research shows that once teachers have joined a union they rarely change allegiance), and their ability to put a positive gloss on the "agreement" in exchange for the kudos of being inside the tent while the NUT—which is still the biggest teaching union—might begin to look irrelevant.

The social partners continued to meet, and a lavishly funded remodeling quasi-nongovernmental organization—the National Remodeling Team—was set up, which excluded those that had refused to sign the agreement—my own union and UCAC—from any further talks. It eventually became clear that officials at every level in the Department of Education had been instructed not to speak to the NUT. Sometimes the other unions also excluded the NUT from local groups, which were discussing the implementation of remodeling

It soon became clear that the NUT was not only being excluded from the talks about remodeling. It was part of my role as president of the NUT in 2004–2005 to engage in discussions with government ministers. As a result of the exclusion of the NUT, I probably was able to do less of this than any previous president. But on one occasion I did meet a minister, a statutory meeting to get the NUT's views on the management allowance structure in schools, an essential part of the pay structure. However, it emerged that the signatories of the workload agreement had already had detailed discussions on this issue and that not only were we not allowed to take part in these discussions, we were not even allowed to see the agendas or the minutes of their meetings. This social partnership group, which was now called the Rewards and Incentives Group (RIG), had already "negotiated" an agreement that would involve many teachers actually losing pay. It initially restricted the top two tiers of the teachers' pay scale to only 20 percent of teachers, awarded on the basis of nebulous performance criteria, and then removed them completely. RIG stated that the new structure would not cost more: "The net national cost of TLR payments will be no greater than the cost of (management) allowances—our expectation is that it should be less." (School Teachers Review Body 2004, 9) Up and down the country teachers on management allowances were told that their head and governing body had a duty to completely break up that structure and start with "a clean sheet." This, of course, gave the green light to bullying heads, of whom there are all too many, to settle old scores and promote their favorites.

Again one has to ask oneself why the other teaching unions should have signed up to such a deal. Why would a self-respecting trade union leader sign up to a cut in members' salaries without even attempting to struggle? When I questioned one of them on why, for example, they had signed up for the concept of safeguarding that would only last three years, when the present system, won in the courts in 1967, gave teachers permanent salary safeguarding, I was told that they were lucky to get anything—the government had originally wanted to give them no safeguarding at all! It is clear that the reason for their quiescence was that they were by now locked into the social partnership agreement and had no wish to disavow it for the narrow sectional reasons mentioned above.

There is one other rather surprising feature of this whole sorry situation that needs to be borne in mind. Teacher unions in the UK have no negotiating rights. They were taken away by the last Conservative administration in contravention of international labor law and have never been restored by the Labor government. So the dialogue going on between some unions and the government under the guise of "social partnership" has nothing to do with negotiations. The discussions start with the premise that there is going to be agreement and are carried out according to the government's agenda. Unless the unions sign up to the "agreement" lock, stock, and barrel—including agreeing to promote it before they have even seen the details—they are simply excluded from the process. The power is all on the side of the government, making it not so much partnership as subordination.

It is perhaps easier for trade unionists from other countries to understand the reasons why union leaders might sell out than to understand the position of the Labor Party, when for many the idea of a party of government founded by the trade union movement is still only an aspiration. An analysis of the development of the Labor Party is beyond the scope of this chapter, but a brief description of its uncomfortable relationship with the trade unions might be helpful. Although it was established by trade unions at the beginning of the last century to represent their interests in parliament, even from the early days it has attempted to cut down on trade union power. In the 1960s the Wilson government came up with a document called "In Place of Strife," which would curtail union rights to take industrial action in exchange for beer and sandwiches with the Prime Minister at Number Ten Downing Street. But this was unsuccessful—the working class at that time was too powerful to be tamed. The next Labor government, in the 1970s, came up with the "social contract"—another attempt to curtail the power of the trade unions. This, too, was largely unsuccessful. However, when Thatcher's government brought in some of the most draconian antiunion laws in the world, making secondary action illegal, bringing in all sorts of complex balloting rules, and incidentally removing the negotiating rights of teachers, the Labor Party in opposition consistently voted against these measures. But in all the years they have been in power, they have not reversed a single one. Indeed, Tony Blair has boasted that the UK has some of the strictest antiunion laws in Europe, making it an attractive place for business to invest.

This contradiction between the founding and the funding of the Labor party and its policies in government are reflected most graphically at the annual Labor Party conference. This annual autumn affair always looks like a meeting of opposites: on the one hand there are the young Oxbridge graduates in their sharp suits with mobile phones glued to their ears, and on the other are the trade unionists who still have block votes at Labor Party conferences concomitant to the political levy that their members pay into Labor party coffers. To temper any possibility of dissent, debate on the conference floor is orchestrated from the chair, reaching its climax with "the leader's" speech. Resistance is stifled—a fine and much publicized example of this happened at the 2005 conference, in which an eighty-two year old man was ejected after shouting "Rubbish!" when the foreign secretary was attempting to justify the government's disastrous Iraq policy. Despite being a member of the Labor Party since 1948, he had his credentials taken away and was even briefly arrested under the "Prevention of Terrorism Act." But largely speaking, the Labor leadership has little to worry about, for most of the leaders of those unions who are affiliated to the party can usually be relied upon to support the leadership if there ever looks to be any seriously threatening opposition from their ranks. Above all, they are anxious to avoid doing anything that they believe might mean "their" Labor Party losing power.

And it is with this government that the teacher trade unions have entered into a "partnership." It is a relationship that reminds one of nothing more than an abusive marriage where one partner is continually vilified and attacked yet unable to leave. In setting out the alternative to this exclusive and abusive relationship it might be salutary to look at the history of teacher trade unions in the UK. And where better to look than at the relationship between teacher trade unions and the last longstanding Labor government in the 1960s?

A History of Struggle

In 1965 it was the then National Association of Schoolmasters (subsequently to become the NAS/UWT) that took action: members of the NAS were instructed to refuse to work with unqualified staff or to take overlarge classes from December 6, 1965. As a member of the NAS executive put it at the time, in a letter to the *Birmingham Post*, "The NAS has determined that a very short term inconvenience to a few parents . . . is more acceptable than the continual damage to the education of part of our child population." (Seifert 1987, 88)

At that time the NAS saw the use of unqualified staff for what it was—a cheap way of avoiding having to deal with teacher shortages caused by low pay and poor working conditions. In the best traditions of teacher trade unionism they appealed to the wider public to see their struggle as one not only for the sake of teachers but to defend the education service for the sake of all children. Although right-wing forces on the NUT executive at that time meant that we were slow to join in with the action, members of the NUT responded enthusiastically. The main actions involved refusing to supervise school meals and refusing to work with unqualified people. As a result of this action two working parties were set up—one led to the abolition of the compulsion to supervise meals and the other to the phasing out of unqualified staff by 1970.

But it was in 1969–1970 that the biggest wave of teacher militancy took place. The industrial action was led in the first instance by the NAS. The NUT executive, which was still the more conservative, dominated as it then was by head teachers, reluctantly accepted a 3.5 percent pay deal—way below the rate of inflation—as part of the then Labor government's prices and incomes policy. This acceptance was overthrown by the Easter conference, for it was becoming increasingly clear that other groups of workers were being awarded much more than 3.5 percent, and teachers were in no mood to be left behind. So this conference "dispersed to its local associations and proceeded to evolve at an extremely rapid rate toward militant trade unionism and strike action" (Seifert 1987, 96). This action rapidly spread all over the country, resulting in 150,000 teachers on strike at any one time between 1969 and 1970. The NUT's discourse started to center on the proposition that teachers had always had to defend themselves and the education service from government attacks and the NUT's newspaper, *The Teacher*, carried front-page articles about previous strikes—significantly, in today's context, the 1923 strike in Camarthenshire for national pay scales. As well as teachers' strikes and sanctions in 1969–1970, significant actions by parents in support of teachers took place. The culmination of all this action was a retreat by the government and a victory for the teacher unions. The day after the deal was reached that gave the teachers £120 of their £135 claim, the NUT executive made a press statement: "The executive regards the interim award as a substantial victory for the Union and a complete vindication of its policy of militant action over the past four months" (Seifert 1987, 103).

It is interesting that the executive had moved it's position on the necessity for taking action as a result of the militancy of its membership. Relentless cutting of budgets and depression of wages causing teacher shortages brought many allies to the teachers' cause—in particular, parents and other trades unions. And it was the tactic of turning to the wider community for support, used both by the NAS and the NUT, that was a key element of the success of the strikes and is one of the many lessons to be drawn from the history of teacher trade union militancy.

But the position in which we find ourselves today is a much more perilous one. We have a government that pays lip service to the right of all children to an education of equal quality while being committed to a neoliberal education agenda—a party in government that boasts of its ability to castrate the very unions that are its main source of income and that treats with breezy disdain the careful research commissioned by the NUT on everything from racism in education to City Academies. And yet this is also a government that is obsessed by public opinion. With the notable exception of the Iraq war, it will not venture a policy until it has consulted its public relations gurus and their focus groups. I have no doubt that if the teacher trade union movement would unite to use its industrial power at the same time as harnessing the huge funds of commitment to education among parents, they would present a force to the government that it would be unable to ignore. So what is standing in the way of this fight back against policies that teachers see overwhelmingly as damaging both to themselves and to education in general?

Learning the Lessons

As teacher trade unionists we are potentially very powerful. While industrial unions have seen their power decline as employers have increasingly been able to close factories and move jobs to areas in the "developing" world where unionization is often weak and wages are low, this is only possible with teachers in a very limited way—for example through the use of "eLearning" in further and higher education. However, the daily job of educating school-aged children cannot be contracted to teachers overseas, nor to a computer program, and the only alternative that the government has found for driving down labor costs has been by allowing unqualified people to teach. For although teaching jobs cannot be exported to other countries, they can be passed to unqualified people—the very agreement that was reached by the so-called social partners in education. So in ceding that principle, the social partnership trades unions have handed the government their most powerful weapon.

There is no doubt also that the division in the teachers' ranks is one of the main breaks to effective action. When struggles have been united they have been effective, as has been shown over and over again, including in this chapter. But the NUT cannot afford to be complacent about its position. Although I am quite convinced that it is right to stand outside the social partnership group, this comes at a cost. There is a real danger that NUT members will feel that the union is being marginalized as a result. Ultimately the only answer to this isolation is for teachers to unite behind one union. There are many determined campaigners for professional unity within my union. Indeed the mantra "Our aim professional unity: one union for all teachers" is printed on all our publications. However there is no prospect of achieving a united union from the top—too many vested interests are at stake. The generosity described by Harold Samuel earlier in this collection, where leaders gave up their positions for the sake of creating a united teaching force in apartheid South Africa, arose from solidarity and struggle against a common enemy. The way forward for the NUT is to unite all those forces fighting against the government's neoliberal agenda, including parents, governors, other trade unionists and teachers from all unions. Only through unity in struggle is there any prospect of meaningful professional unity in England and Wales.

Without a doubt the potential for effective action is still there. Even the threat of a strike on pensions in 2005, which was supported by all the teaching unions, was enough to make the present New Labor government back down on its plans to take serving teachers' pension rights away. There is widespread anger among parents about aspects of the government's choice agenda, about underfunding in many areas, and about aspects of its accountability measures, such as SATs in England. Teachers are angry, too. Almost half are leaving the profession after less than five years in the job, driven out by bureaucracy, stress, and overwork. It is becoming more and more difficult to find head teachers to apply for vacancies—20 percent of schools in England have no permanent head teacher.

As I write, we are in the middle of a struggle against the latest New Labor strategy toward the privatization of education, the Education and Inspections bill, which was so unpalatable to normally quiescent Labor MPs that, despite Labor's majority, it will only be passed with the support of the Conservative opposition. If passed in its present form it will mean, among other things, that all schools could vote to become so-called "trust schools" that would be run by a charitable or religious trust or by the corporate responsibility arm of a private company. Interestingly, when a leading Conservative was asked why he had supported the bill, he said that it was the beginning of enabling legislation to allow the full scale privatization of state schools. Similar plans by the last Conservative government were made largely ineffective by large-scale local opposition from parents and teacher trade unionists. Once again, teaching unions have a historic responsibility to try to overturn this bill. If they fail—which, given the parliamentary arithmetic they almost certainly will—there will be action breaking out all over the country to counter the plans. It will be the job of teacher trade unions to work together to coordinate these campaigns and lead the fight in the defense of free and democratic comprehensive education. And it will be the job of their leaders to develop strategy to use the undoubted industrial power of teachers to defend the education service. They cannot do this if they are locked in the abusive embrace of social partnership, nor if they allow their narrow sectional interests to override their responsibility to serve the interests of all their members and the whole education service. If the teacher trade union leaders would harness the power of their members and turn to their allies in wider society, it would be possible to halt New Labor's determined march toward a neoliberal education system and all that will mean for children and teachers.

As I said at the beginning of this essay, Tony Blair was determined to be a trailblazer for the neoliberal agenda in public services. He is certainly seen in that way by center-right parties in the rest of Europe and valued as an ally by the neoliberals in Washington. I therefore believe that we in the UK teacher unions must be in the forefront of the struggle against those policies, and toward a truly human education service that exists to serve the needs of the people and not the market. The policies of the neoconservatives have been shown to be a disaster in the Middle East. The neoliberal idea that public services and specifically education can be governed by the market is equally disastrous for the children of this planet. It is my hope that I have been able to show the dangers of getting embroiled in social partnership to teacher trade unionists in the rest of the world just as we can learn from their heroic struggles. As for us in the UK, our historic responsibility is to learn the lessons from our own militant past and start the long fight back.

References

Blair, T. 1999. Speech to Venture Capitalists Association, London, England, July 6, 1999.

————. 2004. Speech, Evidence to Select Committee on Liaison. Questions 180–99, London, England, July 6, 2004.

Department for Education and Skills. 2003. *Time for standards: Guidance accompanying the Section 133 regulations, Education Act 2002*. London: Department for Education and Skills.

Kelly, J. E. 2004. Social partnership agreements in Britain: Labor cooperation and compliance. *Industrial Relations: A Journal of Economy and Society* 43 (1): 267–92.

Hayes, D. 2004. WI to the rescue: Plan to cut teachers' hours by drafting in Women's Institute and caretakers. *London Evening Standard* (England), November 20.

National Union of Teachers. 2003. The DfES Paper you weren't supposed to see: *Workforce reform. Blue Skies*, 2.

Seifert, R. V. 1987. *Teacher militancy: A history of teacher strikes 1896–1987*. London: Falmer.

School Teachers Review Body. 2004. Evidence from the Rewards and Incentives Group: September 2004 Agreement on Teachers' Pay.

United Nations. 1948. *Universal Declaration of Human Rights*. New York: United Nations.

CHAPTER 27

Building the International Movement We Need
Why a Consistent Defense of Democracy and Equality Is Essential

Lois Weiner

As a researcher, I study teachers' work and their unions. However, my most valuable education on the subject occurred when I began my career as a teacher in a California secondary school more than thirty-five years ago. I was recruited to the local of the American Federation of Teachers (AFT) by another teacher I knew from my involvement in anti–Vietnam war activity. At my first union meeting I heard—and voted on—the "plan of action" presenting union members' negotiating demands on issues ranging from how curricula were determined to class size maximums, as well as salary and medical benefits. Even before the first meeting of the local, another union member in my school welcomed me—and asked me to help by putting informational material in teachers' mailboxes.

We were a minority organization (the affiliate of the NEA, the National Education Association, had more members than we did), and we were not homogeneous in terms of our political ideals nor our ideas about pedagogy. Some members, mostly older men who taught teenagers (in U.S. secondary schools), held traditional ideas about instruction and curriculum, as well as what is now called "the culture wars." Others, myself included, identified strongly with the social movements pressing for full equality, for racial minorities, women, immigrants, and gays. Usually those of us who were younger saw union work as an extension of other political activity. As was true of teacher union activists elsewhere in the world, many of us came from the 1960s student movement and were attracted to pedagogical ideas John Dewey had promoted a century before, though we did not realize they were his. Dewey was a philosopher who helped found the AFT in the late nineteenth century and whose slogan, "Democracy in education; education for democracy," years later graced the masthead of AFT publications when classroom teachers again began to organize in U.S. cities, demanding the right to form unions that represented them.

Our deep and often contentious disagreements were salved by a commitment to make the union a vehicle to defend our dignity and rights as teachers and school workers. We knew our ideas and actions counted because we could influence the union's exercise of *institutional* power and the union could project our *collective voice and power*. These characteristics are what make neoliberal politicians and pundits accurately see—sometimes better than do teachers themselves—teacher unions as a threat to neoliberalism's project. Union principles of collective action and solidarity contradict neoliberalism's core principles, reification of individual effort and competition.

Individuals speak for organizations, but it is not individual goals or personalities that create conflict between unions and governments. Rather, unions, when they are doing what they should, encourage notions and practices that run counter to neoliberalism's central tenets. Unions develop because individuals have limited power when they confront the employer at the workplace and in society as a whole. Unions are based on the notion that when people who earn a salary join together with others at the workplace and decide what is most important and how best to advance the goals they hold in common, they bring a strength much greater than the sum of their parts. Another reason that teacher unions pose a serious threat to the neoliberal project is that they are a stable force. Unions are organizations with institutional roots. A union remains through changes of personnel on both sides of the bargaining table. A union is able to draw on a regular source of income from its members, and this provides teacher unions with an organizational capacity that is seldom acquired by groups formed on single issues, like peace or global warming or gay rights. Unions are the single most powerful threat to neoliberalism's exercise of unchecked power, which explains the constant barrage of antiunion propaganda in the media, often emanating from reports issued by seemingly objective "watchdog" organizations or commissions.

However, the very factors that create teacher unions' potential and actual strength also create problems. Unlike social movements, which bring together people who are in agreement on an issue, like the despoiling of the environment, unions are membership organizations that are obligated to serve all members, regardless of their beliefs. Further, the institutional arrangements that give unions stability and strength, like the automatic collection of dues from members' paychecks and its officers' acquisition of specialized knowledge and skills to enforce complex provisions of negotiated settlements, also make the unions prone to bureaucracy.

Neither unions as organizations nor union members as individuals are immune to prejudices that infect a society, even when these prejudices contradict the union's premises of equality in the workplace and solidarity. Unions, like classrooms, are affected by social, economic, and political life. Teacher unions are buoyed by successful, widespread activism about politically progressive causes and weakened when progressive movements are on the decline. When I started teaching, it was possible for a union to function bureaucratically, without mobilizing members, and still make economic gains. Now, neoliberalism's power, even apart from its assault on public services and education, threatens the very existence of the unions. Moreover, whereas labor unions could previously operate effectively within national or local borders, today unions must mount a global response because educational policies, capital, and jobs rapidly move from one nation to another, as Larry Kuehn and Susan Robertson explain in their chapters.

In the rest of my essay I explain what I think is probably the most essential principle for creation of a global movement capable of stemming and ultimately turning back the neoliberal assault: The consistent defense of democracy and social justice within the teacher union movement itself and throughout the world.

Neoliberalism and Inequality

Neoliberalism's power, its ability to overturn so many long-standing regulations, practices, and beliefs, emerges from its supporters' willingness to employ all means at their disposal, including, when needed, military and paramilitary force. But its strength is also based on the appeal of its rhetoric and its ideology. Neoliberalism advances its policies as the best and only way to bring economic prosperity to those who are in poverty, and while it is tempting to ignore its stance of ameliorating inequality, this is a flawed strategy (Weiner 2005). Teacher unions need to contrast neoliberalism's professed concern about poor people with the reality of what occurs when its policies are implemented. We have much evidence that policies enacted since World War II aimed at equalizing achievement had some success, as was hoped. However, they also failed to eliminate inequality in school outcomes, partly because of the way the reforms were conceptualized and partly because they were *too limited* to address the deeply rooted causes of academic failure, causes both in schooling and in the society (Bastian et al. 1986).

When special educational programs for "disadvantaged" youth were created, they were based on a "compensatory model" subsequently used for other groups, like students with disabilities and immigrants requiring instruction in the school language. From the start this model had serious political and educational limitations, including the way "success" was measured, through students' scores on standardized tests (Bastian et al.1986). Thus, neoliberalism's current use of standardized tests to control curricular outcomes was first made possible because reformers, including U.S. teacher unions, accepted standardized testing (Weiner 2005). Another flaw in the compensatory model was the implicit acceptance of its educational theory about why some children do not succeed in school, that children of historically oppressed groups are under-achieving because they have problems, deficits, that needed remediation. Compensatory programs in the United States provided much-needed but relatively small funding increases for the education of minority students while ignoring systemic inequalities resulting from the structure and organization of schooling. School improvement became a substitute for concentrated attention to social, political, and economic inequality outside the school walls. In contrast to the compensatory model, Brazil's Escola Plural Program, which Álvaro Hypolito describes, locates school reforms within a broad political program to alleviate inequality. In contrast to compensatory or remedial programs, the Escola Plural Program has attempted to reorganize the structures and organization of schooling, including the way learning is evaluated.

To advance a credible argument about why neoliberalism is destructive, teacher unions need to acknowledge educational inequality's persistence and its eradication as a political and social priority. However, the case also needs to be made that school reforms can ameliorate inequality but not, by themselves, resolve economic and social inequality and that neoliberalism's economic and political policies increase school dysfunction and academic failure for children who are most vulnerable (Lipman 2000).

Teacher unions must often draw a clear line in the sand between neoliberal ideology and a contrasting vision of society. It may seem more practical to compromise, as did teacher unions when they first endorsed standardized testing as a fair, effective assessment tool and compensatory programs as providing much-needed funding to schools. In reality, these bargains undercut development of alliances with parents and citizens. The immediate economic and political rewards offered by neoliberal politicians in return for collaboration in their project cannot compensate for the disorientation this partnership generates about alternative social arrangements. Success in turning back the neoliberal reforms depends on teacher unions educating a new generation of teachers and activists about public education's economic and political responsibilities to equalize social relations. As Mary Compton explains, standing outside the partnership causes hardship, but collaboration means forfeit of critique—and of the possibility of cohering a countervailing movement.

Rhetoric and Reality

Although much neoliberal rhetoric focuses on the need for "higher standards" that arise from the creation of a global economy, World Bank documents describe quite a different blueprint for the world's poorest countries (Weiner 2005). The World Bank presumes that poor countries can become prosperous only if they compete for low-paying jobs that require no more than a basic education, jobs that transnational corporations can easily move to a country whose workers are paid less than prevailing wages in a competing nation. When workers have jobs that require little education, schools will not be teaching most students high-level content and skills, so poor nations do not require teachers who are themselves well educated or highly skilled. As Ken Zeichner explains, the assumption driving neoliberal reforms is that children from poor and working class families need only teachers who are "good enough" to follow scripted curricula. Teachers who have a significant amount of education are costly to employ and so impede neoliberalism's goal of slashing expenditures for all public services, including education. As in so many other of its reforms, neoliberalism's alterations in teacher education have actually exacerbated educational inequality by systematizing dual tracks of teachers, one stable and well prepared for prosperous communities, the other minimally trained and transient for schools serving poor, rural, and working class families (Gaynor 1998).

The massive project of social re-engineering undertaken by neoliberalism pressures nations to replace career teachers with those who have only minimal training without evidence that this strategy actually results in educational gains. The following excerpt from a draft report issued by the World Bank makes explicit the economic rationale for replacing experienced teachers. The draft was heavily criticized, and this segment was subsequently deleted from the final version of the report in a process I document elsewhere (Weiner 2005):

> With their political power, teachers and doctors are able to protect their incomes when there is pressure for budget cuts. The only thing left to cut, therefore, is non-wage operations and maintenance expenditures. Many governments

have responded by creating a second class of teachers who are outside the civil service, and are correspondingly paid less with fewer benefits. The experience in several West African countries shows that there are many people willing to take these jobs (a recent announcement in Senegal generated 30,000 applicants for 1,000 positions); even if they are less qualified, the evidence on student performance is mixed; and, over time, these contractual workers have come to dominate the public service, as in Benin (Devarajan and Reinikka 2002, 6).

The expression of naked intent to destroy public service unions, especially those of teachers, came as a surprise to many teacher union officials in the global north, where neoliberal rhetoric about eliminating inequality, leaving no child behind, and the need for all children to have "highly qualified teachers" has masked its real agenda (Weiner 2007). Nonetheless, the idea that the government has no obligation to provide an advanced education has been introduced in the United States, as in the decision of a New York state appeals court that students are legally entitled only to an education that prepares them for a menial job and jury duty. This court decision overturned a judgment that would have given New York City schools more funding (Perez-Pena 2002).

In much of the world, neoliberal policies in education have been demanded by the World Bank and the International Monetary Fund (IMF) as preconditions for economic aid. Namibia's experience, described by John Nyambe, typifies how developing countries have been forced to accept restructuring of their educational systems. But how can we explain the acceptance of neoliberal reforms in developed nations? Neoliberalism's exploitation of the rhetoric of progressive reform is one factor. Another is neoliberalism's strategic alliance with political camps on what was once called the far right, including religious fundamentalists and neoconservatives. Neoconservatives, like social and religious conservatives, aim to overturn legal and cultural gains won by social movements starting in the 1960s, like the right to abortion and affirmative action or "quotas" in hiring. However, unlike some on the religious right, neoconservatives ardently support U.S. political and military hegemony, exercised through direct interventions and the control of client states. Because neoconservatives identify "democracy" as being "Western" and see "democracy" and "capitalism" as fused, they are often close allies of neoliberalism. As a U.S. think tank that identifies itself as "free market" notes: "The Mackinac Center for Public Policy . . . does not address issues that are primarily questions of social ethics such as abortion, censorship, and gambling. While these are important issues, the Mackinac Center has adopted an inclusive strategy. By gaining the support of all who recognize the importance of sound economic policy—despite their disagreement over social issues—we are able to more effectively accomplish our key objective: to establish a more sophisticated level of political and economic understanding among Michigan citizens and decision makers" (Mackinac Center for Public Policy, n.d.).

The focus of this essay does not permit full discussion of how neoliberal ideas differ from others on the right, but as Susan Robertson explains in her essay, policies advanced by neoliberals that claim to be based on the "free market," such as those endorsed by the Mackinac Center, often rely on government interventions and financial support. Álvaro Hypolito notes why these ideological differences are significant. Neoliberalism's willingness to partner with social conservatives (and vice-versa) is a thorny challenge for teacher unions because some union members may share the social attitudes and political beliefs of social conservatives. For this reason a union's commitment to democratic norms is essential so members—who, as Nina Bascia explains, differ in what they want from the union—view its policies as legitimate. On the other hand, the role of a union leadership is not only to safeguard democratic norms of debate and decision making but also to set out the principles of social justice that should inform its policies. It is the mobilization of members in defense of social justice issues—rather than only the passage of resolutions—that is the hallmark of a union that has adopted the strategy I recommend. Harold Samuels' recommendations about the union's future in South Africa, Chris Stewart's reflections on Aboriginal education in the British Columbia Teachers' Federation, and Rob Durbridge's analysis of how the Australian union has advanced a social justice agenda illustrate what this principle can look like in real life.

One ideal of social justice that teacher unions have a special stake in pursuing is gender equity. A problem raised by Kathleen Murphey, Nina Bascia and other contributors to this book is the fault line among teacher organizations about teacher professionalism: Are teachers professionals or workers? I suggest that the way this question is most often posed ignores gender inequality and with it the gendered nature of teachers' work. Although I cannot discuss this issue as fully as the topic deserves, considering gender may help address conflict between the "industrial union" and the "professional model" for teacher organizations alluded to in several essays. Both models ignore that teaching, in contrast to a career in law, for example, shares with mothering characteristics of work that occurs in the family setting (Biklen 1995; Freedman 1990; Weiner 2003). In most societies the work that mothers and teachers do, caring for children, is inadequately valued, as is demonstrated by the struggles for adequate funding of schools even in prosperous times. The "parent" who most often visits school and superintends the child's education is a mother, so the interaction between parents and teachers is primarily one of women working with women (Biklen 1995). Alliances between teachers, teacher unions, and parents will likely be comprised in the main of women working together.

But what about the class and social divisions among women? Aren't privileged mothers the natural allies of neoliberalism? Writing about mothers' involvement in their children's education, British sociologist Diane Reay (1998) argues that white and middle-class parents are not homogenous in the way they view schooling, although there is a "vociferous minority" (Reay 1998, 129) who advance their own children's interests at the expense of those with less power and status. While it is important not to romanticize how gender may mediate class, it is also accurate that mothers, as the parent who most often must navigate between home and school, bear the brunt of the additional family stress that accompanies neoliberal policies that stratify education.

While gender "counts" in the way that teachers understand their work— and their commitments to unions—so do race, class, and histories of oppression. One way to define teacher "professionalism" that addresses the gendered nature of teachers' work and complements unions' social justice ideals is to understand that as professionals and workers, teachers serve communities. This revised definition of professionalism will appeal to teachers who understand their work as serving *particular* children and families, a factor that is especially important in winning over to teacher unionism teachers and communities whose personal experience of unions has been primarily that of social exclusion, as with African Americans in the United States (Golin 2002). Neoliberal rhetoric about educational inequality has a particular resonance for those who have suffered discrimination by unions directly, or as a result of employment policies that unions have accepted. They do not easily accept the argument that labor unions are a natural ally of struggles for equality and justice. Often young teachers who are activists in social justice struggles and those who see their teaching commitments as the best way to improve the society view the union as irrelevant—or worse, as an opponent. For reasons both practical and principled, teacher unions need to involve members in social justice struggles that are not directly germane to the "bread and butter" of contracts and negotiations. Admittedly, support for social justice struggles pits teacher unions against powerful enemies. And the union will likely face debates and criticisms from members who want the union to limit itself to "practical" concerns, as Chris Stewart explains elsewhere in this volume. However, if the union makes debate and decision making democratic *and* mobilizes members to win material gains, devoting resources to social justice campaigns makes a priceless internal contribution, recruiting and educating a new generation of union activists, as well as winning new external allies.

A Consistent Defense of Democracy

Our ability to build an international movement to reverse policies that are destroying public education (as well as the quality of life generally, working conditions, the environment, etc.) depends on projecting a vision of human emancipation, a world that provides both political freedom *and* social control over economic resources. To do this, we should be clear about what is wrong with neoliberalism's definition of "democracy." Barber (2004) notes:

> Neoliberalism's two ascendant principles are the priority of markets (government is part of the problem rather than the solution while markets can solve the problems government creates) and the substitution of consumers for citizens (democracy is defined less by common public choice than by private market decisions). . . . With these principles in place, it is easy to understand how neoliberals come to believe that democracy itself is the same thing as market democracy and the spread of markets and the ideology of privatization associated with it is the same thing as the spread of democracy. In Russia in 1989 after the collapse of the Soviet regime, for example, it was easy for many Westerners to think that privatizing communism's state industries was tantamount to establishing a democratic regime. More recently in Iraq, proconsul Paul Bremer announced that Iraq would have a privatized energy and industrial sector as well as market-dominated media outlets in the apparent conviction that this was same thing as establishing the foundations for democracy. (n.p.)

Neoliberalism makes the "market" synonymous with democracy, and refuting this equation is key to winning public opinion in campaigns against key structural reforms, especially privatization of services and elimination of control by local education authorities (school boards in the United States). Unfortunately, U.S. labor unions, including those representing teachers, the National Education Association (NEA) and AFT, often accept neoliberalism's definition of "democracy" because they see the world refracted through the eyes of U.S. capitalism, which dominates the World Bank and other international financial organizations. Teachers in the United States have been able to wear blinders about neoliberalism's impact because unlike their equivalents in the countries receiving the World Bank "largesse," U.S. teachers have been relatively protected from the program the United States enforces on the rest of the world. Still, neoliberalism has "come home" to the United States in the form of the federal package of school policies called No Child Left Behind

(Weiner 2007). Opposition to NCLB has combined with growing disillusionment with the war in Iraq, now officially opposed by both NEA and AFT, to widen the political gulf between U.S. teacher unions and the Bush administration.

Evidence of AFT's flawed definition of democracy is apparent in the online version of the printed publication sent to its 1.4 million members (AFT 2003a) each month. One article asks teachers to "display freedom's spread" in their classrooms by using a survey produced by Freedom House, a nonprofit organization long associated with neoconservative political ideas (Barahona 2007). Unfortunately, AFT's use of the Freedom House materials is evidence that AFT's worldview continues to be dominated by ideas held by Albert Shanker and Sandra Feldman, who controlled the AFT national organization as presidents, (1974–1997 and 1997–2004, respectively). They exercised considerable influence internationally, primarily because of their close personal and political ties with powerful neoconservatives in the United States and their desire to be partners in U.S. capitalism's global designs (Weiner 2005).

Freedom House annually produces a map of the world divided into societies that are "free," "unfree," and "partly free" according to its analysis of access to democratic rights. The questions it uses to gauge democracy's presence include free and fair elections, the right to organize into political parties independent of the government, existence of free trade unions, academic freedom that excludes political indoctrination, and civilian control of police. All of these are political freedoms teacher unions need to endorse, unequivocally. But Freedom House also questions: "Are property rights secure? Do citizens have the right to establish private businesses? Is private business activity unduly influenced by government officials, the security forces, or organized crime?" (AFT 2003a; Freedom House 2004). The Freedom House map indicates that answers to this last question seem to "trump" all other rights (Freedom House 2004).

Limitations of political freedom in capitalist democracies, the United States included, and states that the U.S. government supports are minimized or ignored. Violations of democracy in countries the United States considers its foes are closely monitored. For example, the Freedom House analysis about U.S. press freedom in 2003 (Freedom House 2004) contradicts conclusions of International PEN, an organization of writers who take up violations of free speech, in its report "Anti-Terrorism, Writers and Freedom of Expression—A PEN Report" (Whyatt 2003). Freedom House observes that "freedom of expression is guaranteed by the constitution" and also that governmental "restrictions on domestic press coverage, begun after the September

11, 2001, terrorist attacks, were expanded in preparation for U.S. military action in Iraq" (Freedom House 2004). Freedom House explains each restriction imposed by the government on the press with the official rationale. In contrast, the International PEN report (Whyatt 2003) warns about the peril for freedom of expression in the United States, as well as for other countries that receive the green coloration of "free" on the Freedom House map. Conditions include "greater fear among citizens, especially in the West, of terrorist attacks, especially at the hands of those deemed to be 'Islamic Fundamentalists'; a rush by various governments to grant their intelligence agencies and their police new powers aimed at identifying and apprehending terrorists; two major invasions—of Afghanistan and Iraq—by a coalition of forces led by the United States; a continuing 'war on terrorism' in which the next stages are unclear" (10).

PEN's American Center notes in its Campaign for Core Freedoms that the United States has "weakened the power of the American people to monitor government activities and built barriers that restrict the flow of information and ideas; and created a shadow legal system that undercuts basic due-process protections" (PEN 2003). But in an open letter the AFT circulated and reprinted in its magazine (AFT 2003b), communism and a "new tyranny— Islamic extremism" are cast as dwarfing all other threats to democracy. Further evidence of the AFT leadership's lopsided view of threats to democracy is found in a review of its publications and press releases. The AFT website contains only two instances in which the term *neoliberal* is used, and both are in materials for its college and university unit, written in 2001. Communism, on the other hand, is mentioned in twenty-three documents, Cuba in twenty-one, and China in 215. Mention of Israel is found in forty speeches, resolutions, and articles, almost all of which mirror neoconservative positions on Israel and do not criticize, even mildly, Israel's treatment of Palestinians. No mention is made about the discriminatory treatment of Arabs in Israeli textbooks described by Nurit Peled-Elhanan in this volume. Palestine is mentioned three times, each time negatively. Still, the AFT is sometimes compelled to criticize a government that is friendly to the United States, generally on the basis of a "bread and butter" issue. And so despite Washington's friendly relations with the government of Mexico, AFT issued a public statement condemning attacks on teachers in Oaxaca. And while the AFT national office stands firmly in the neoconservative camp in terms of Israel's treatment of Palestinians, in a May 2005 letter the union states its opposition to the government's firing of 6000 Israeli teachers (AFT 2005).

NEA and AFT are the largest constituent members of Educational International (EI), the merged international confederation of teacher unions that was formed after the fall of the Berlin wall and collapse of the Communist bloc. Until recently AFT and NEA dominated EI, run by staff and officers in Brussels who focused its energies on admirable advocacy work for individual teacher unionists. As EI 's then general secretary observed in a speech in 2001, one of its strengths is that it is a membership organization; yet at the same time, the confederation depends on 2000 union officers and activists to speak for more than 24 million members (cf. Weiner 2005). Education International has experienced a revitalization with the election of a new president, Thulas Nxesi, who is leading the organization in a different political direction, one that confronts neoliberalism's global assault on teacher unions and public education.

EI's rejection of the AFT-NEA political worldview blinded to neoliberalism's dangers is essential. Just as necessary is a consistent defense of democratic rights, one that does not overlook suppression of political freedom or human rights carried out by governments or movements that are opposed to neoliberal policies. When teacher unions fail to support a people's right to the fullest political freedoms, ones exercised (though increasingly with restraints) in liberal capitalist democracies, unions inadvertently reinforce neoliberalism's bogus claim that its project alone offers a way to have democracy. Refuting neoliberalism's contention that "democracy" is synonymous with the "free" market is essential to undercutting the claim repeated constantly in the media and from governments that capitalism insures political freedom, that the only alternative to corrupt, capitalist plutocracies is an undemocratic society. Progressive activists unwittingly reinforce neoliberalism's contention that "there is no alternative," no way to organize society to provide the fullest human freedom, when they excuse violations of democratic rights in governments and popular movements that resist neoliberal policies or use the rhetoric of opposition. María de la Luz Arriaga Lemus explains why the struggle for democracy is essential in Mexico. I would add that the same is true internationally.

What does this consistent defense of democracy look like? One illustration is the work done by the Campaign for Peace and Democracy (CPD) in defense of an imprisoned Iraqi union leader:

> We, who opposed the US-led war on Iraq and who call for an immediate end to the occupation of that country, are appalled by the torture and assassination in Baghdad on January 4, 2005, of Hadi Salih, International Officer of the Iraqi Federation of Trade Unions (IFTU) . . . though we disagree strongly with the IFTU's support of UN Resolution 1546, which supports the US military presence in Iraq. This resolution has been used by the Bush Administration to justify keeping US troops in the country.

> We also oppose the victory of those elements of the resistance whose agenda is to impose a repressive, authoritarian regime on the Iraqi people, whether that regime is Baathist or theocratic-fundamentalist. . . . The continuing occupation of Iraq, which grows more brutal with every passing day, only strengthens these elements, increases their influence over the resistance and makes their ultimate victory more likely. We further oppose the occupation because it is part and parcel of an imperial US foreign policy that shores up undemocratic regimes like those of Saudi Arabia and Egypt, gives one-sided support to Israel against the Palestinians, and promotes unjust, inequitable economic policies throughout the world. Not only in Iraq but throughout the Middle East and globally US foreign and military policy either directly or indirectly subverts freedom and democracy (CPD 2005).

In the past CPD has organized similar campaigns, for example joining a condemnation of U.S. threats to Cuba with a protest against Cuba's arrest of scores of people for nonviolent political activity. Being adamant that teacher unionism internationally defends both economic and political rights everywhere gives the lie to the neoliberal claim that only the "free" market can protect freedom. We greatly undercut neoliberalism's hypocritical claim that teacher unions act only in their "selfish" interest, that is, on the kind of "bread and butter" issues on which the AFT will criticize a government friendly to the United States. The contents of this book demonstrate the global nature of neoliberalism's assault, but authors also suggest the contour of a teacher union–led movement that can turn back the attack. From China to India, from Namibia to Denmark, from South Africa to Mexico, teachers want unions that are independent of the government, unions that defend public education and the dignity of teachers and teaching. We see the outline of an international movement in the election of EI's new president and of EI's support to teachers who are jailed or kidnapped when they assert their rights to form unions, as well as in the development of an international alliance of teacher unions in the Americas to resist inclusion of education in trade treaties. The emergence of the IDEA network, described by Larry Kuehn, illustrates how struggles to democratize unions locally contribute to the growth of an international movement. The IDEA network's first U.S. participants, University of New York's Professional Staff Congress and the United Teachers of Los Angeles, are headed by leaders elected through victories of rank-and-file groups pledged to democracy and social justice. An international teacher union movement, one that works with global justice activists outside its ranks, can build successful resistance to neoliberalism. But this movement will only be built by people who believe that we have the capacity to develop alternatives and do not have to accept the existing, flawed arrangements that allow elites to control our lives.

Schools remain essential sites of resistance to neoliberalism. Struggles to protect public education are key in holding the democratic state to account. Teachers and teacher unions have a key strategic role in organizing the alliances that can push back the neoliberal offensive and win for children of the world the futures they deserve. For the movement to have the moral authority it must, and for people to have the confidence that they can remake the world in ways that those who are now in control say are impossible, an international movement of teachers needs to transform its unions, to make them consistent defenders of equality, social justice, and democracy—that is, of human emancipation.

References

American Federation of Teachers. 2003a. American educator fall 2003 notebook. AFT Publications. http://www.aft.org/pubs-reports/american_educator/fall2003/notebook.html (accessed July 12, 2007).

———. 2003b. Education for democracy: A statement signed by over one hundred distinguished leaders. *American Educator*, 6–23. http://www.aft.org/pubs-reports/american _educator/fall2003/Democracy.pdf (accessed July 12, 2007.).

———. 2005. *Letter to the Histadrut.* http://www.aft.org/topics/international/downloads/ Histradrut052005.pdf (accessed July 7, 2007).

Barahona, D. 2007. The Freedom House files. *MRzine*, March 1. http://mrzine.monthly review.org/barahona030107.html (July 12, 2007).

Barber, B. R. 2004. Taking the public out of education. *School Administrator*, 61:10–13.

Bastian, A., N. Fruchter, M. Gittell, C. Greer, and K. Haskins, K. 1986. *Choosing equality. The case for democratic schooling.* Philadelphia: Temple University Press.

Biklen, S. K. 1995. *School work: Gender and the cultural construction of teaching.* New York: Teachers College Press.

Campaign for Peace and Democracy. 2005. Opponents of the occupation condemn attacks on Iraqi trade unionists. http://www.cpdweb.org/ (accessed July 7, 2007).

Devarajan, S., and R. Reinikka. (2002). Making services work for poor people, 25 July 2002. http://siteresources.worldbank.org/INTWDR2004/Resources/17976_Reinikka ShantaInitialFramework.pdf (accessed November 14, 2007). Freedom House. 2004. *Map of press freedom 2004 edition.* http://www.freedomhouse.org/template.cfm?page=251&year =2004 (accessed July 12, 2007).

Freedman, S. 1990. Weeding woman out of "woman's true profession": The effects of the reforms on teaching and teachers. In *Changing education: Women as radicals and conservators*, ed. J. Antler and S. K. Biklen, 239–56. Albany: SUNY.

Gaynor, C. 1998. Decentralization of education: Teacher management. In *Directions in development.* Washington DC: World Bank. http://www-wds.worldbank.org/servlet/WDSContent Server/WDSP/IB/1998/02/01/000009265_3980429110720/Rendered/PDF/multi_page .pdf (June 24, 2004).

Golin, S. 2002. *The Newark teacher strikes: Hopes on the line.* New Brunswick, NJ: Rutgers University Press.

Mackinac Center for Public Policy. n.d. *Is the Mackinac Center for Public Policy liberal? Libertarian? Conservative?* http://www.mackinac.org/article.aspx?ID=1663 (accessed January 8, 2007).

Lipman, P. 2000. Bush's education plan, globalization, and the politics of race. *Cultural Logic* 4 (1). http://eserver.org/clogic/4-1/lipman.html (accessed January 4, 2002).

PEN American Center. 2004. *The campaign for core freedoms.* http://www.pen.org/page.php/prmID/293 (July 12, 2007).

Perez-Pena, R. 2002. Court reverses finance ruling on city schools. *New York Times*, June 26, A1.

Reay, D. 1998. *Class work: Mothers' involvement in their children's primary schooling.* Bristol, PA: UCL Press.

Weiner, L. 2003. Class, gender, and race in the Newark teacher strikes. *New Politics* 9 (2). http://www.wpunj.edu/~newpol/issue34/weiner34.htm (accessed July 7, 2007).

———. 2005. Neoliberalism, teacher unionism, and the future of public education. *New Politics* 10 (2): 101–12. http://www.wpunj.edu/~newpol/issue38/Weiner38.htm (accessed July 7. 2007).

———. 2007. NCLB, U.S. education, and the World Bank: Neoliberalism comes home. In *Facing accountability in education: Democracy and equity at risk*, ed. C. E. Sleeter, 159–71. New York: Teachers College Press.

Whyatt, S. 2003. *Anti-Terrorism, Writers and Freedom of Expression—A Pen report.* http://www.internationalpen.org.uk/images/article/Anti-terrorism,%20writers%20and%20freedom%20of%20expression%20(English).pdf (accessed July 12, 2007).

World Bank. 2003. *World development report 2004: Making services work for poor people.* http://econ.worldbank.org/wdr/wdr2004/text-30023/ (accessed June 25, 2004).

Vignette 27.1

Editors' note: Resistance to neoliberal policy often occurs without teachers realizing the extent of international struggle against the attacks. Below we note only a fraction of the actions teachers have taken in defense of their jobs and public education, from January through June 2007. The reports on strikes in North America were compiled and written by Philip Hayes. Mary Compton provided the other listings.

China

January 1, 2007

Teachers in the Chinese town of Guyangzhou come out on strike to demand higher wages.

Angola

February 22, 2007

Angolan members of the General Assembly of Teachers go on national strike to demand changes to their official status and timely payment of salary despite threats and physical harassment by security forces.

India

February 26–27, 2007

Fifty thousand primary school teachers take part in a mass rally in New Delhi to protest aspects of the government's education bill. Their protest emphasizes the right of a child to free, quality, and equitable education.

Slovenia

March 15, 2007

The teachers union ESTUS demonstrates to defend the public education system against government privatization plans.

UK

April 7, 2007

Delegates at the annual conference of the National Union of Teachers (NUT) vote to ballot for strike action over pay.

Senegal

April 11–26, 2007.

Teachers strike for three weeks for increased housing and research allowances and against attempts to increase their work hours.

Argentina

April 16, 2007

Argentinian teachers are joined for one hour by bank and public employees in their nationwide strike today to protest the killing by security forces of a chemistry teacher, 41-year-old Carlos Fuentealba, during a teachers' pay protest.

Mayotte

April 16, 2007

Primary teachers are on strike for three weeks demanding improvements to the education system in the island.

Niger

April 18, 2007

Teachers in the town of Niamey strike, demanding pay increases. (Teachers in Niger are paid between $960 and $1400 per year.)

Iran

April 29, 2007

Teachers throughout Iran hold mass protests and strikes against low wages despite violent oppression by the government.

Trinidad and Tobago

May 26, 2007

Trinidad and Tobago Unified Teachers Association (TTUTA) leaders lead a mass demonstration of teachers demanding a salary increase.

Poland

May 29, 2007

Nearly half of Polish schools take part in a national strike to demand a 20 percent pay raise for education staff, more education spending, and upholding of the right to early retirement.

South Africa

June 1, 2007

Up to one million South African public sector workers, including school principals and teachers, go on strike demanding higher pay.

Peru

June 14, 2007

Members of the teacher union SUTEP go on strike against the government's failure to keep to agreements.

Canada

March 14, 2007

Laval University's SCCCUL members go on indefinite strike, calling for higher wages and better working conditions. The teachers on strike, known as chargés de cours, are part-time teachers who are hired by the university to teach specific lessons.

United States

January 29–February 11, 2007, Lee County, Arkansas

The strike for over 150 teachers and additional support personnel of the Lee County Education Association lasts for two weeks.

March 13–25, 2007, Philadelphia, Pennsylvania

Four hundred professors at the Community College of Philadelphia in Local 2026 of the American Federation of Teachers strike for twelve days, canceling classes for 37,000 students.

April 5–27, 2007, Hayward, California

Members of the Hayward Education Association strike for ten days. Media-savvy teachers in the HEA use YouTube, Wikipedia, and blogs to communicate with over 1,300 striking teachers through the Internet.

In addition, four strikes of affiliates of the National Education Association occur in school districts (local education authorities) in Pennsylvania in May and June.

About the Authors

María de la Luz Arriaga Lemus is currently on the faculty of the Universidad Nacional Autónoma de México, Facultad de Economía. She has been a teacher educator and a teacher of political science and economics and participates in the Tri-National Coalition in Defense of Public Education and Red SEPA (Red Social para la Educación Pública en las Américas).

Nina Bascia is Professor, of Theory and Policy Studies in Education at the Ontario Institute of Studies in Education, University of Toronto, Canada. She wrote *Unions in Teachers' Professional Lives* (1994) and is co-editor of *The Sharp Edge of Educational Change* (2000) and *The International Handbook of Educational Policy* (2005).

Eberhard Brandt is an executive member of the Gewerkschaft Erziehung und Wissenschaft (GEW) from the state of Lower Saxony, Germany.

Basanti D. Charkraborty is on the faculty of the Early Childhood Education Department at New Jersey City University. She received her PhD in Education from Utkal University, India. Her book, *Education of the Creative Children*, about the effectiveness of the Indian government's National Rural Talent Scholarship scheme, was published by the Indian Council of Social Sciences Research (ICSSR).

Yihuai Cai is a dual-title doctoral student in Curriculum and Instruction (Language and Literacy Education) and Women's Studies at the Pennsylvania State University. Her research focuses on gender and education, feminist pedagogy, and feminist poststructuralist ethnography. She is currently on fieldwork in a rural village in central China conducting an ethnography on girls' education and the changes in rural communities.

Marguerite Cummins Williams was a teacher in Canada and Barbados until her retirement in 2002. She is a former President of the Barbados Secondary Teachers' Union and was an Executive Board member of Education International from 1993 to 2004. She now serves on the EI Committee of Experts.

Urban Dolor is a former President of the Saint Lucia Teachers' Union and former vice-president of the Caribbean Union of Teachers. He is currently Vice Principal of the Sir Arthur Lewis Community College.

Susanne Gondermann is chair of the advisory committee on comprehensive schools of the Gewerkschaft Erziehung und Wissenschaft (GEW), Germany.

Rob Durbridge is the Federal Industrial Officer and former Federal Secretary of the Australian Education Union (AEU).

Elaine Hampton, Associate Professor in Curriculum and Instruction, is Chair for the Department of Teacher Education at the University of Texas in El Paso. She teaches field-based science methods classes, as well as curriculum studies, educational research, and assessment. Her research examines the educational programs in the border area and their impact on border learners.

Philip Hayes is a longtime classroom teacher of social studies at Brookhaven High School in Columbus, Ohio. He is a current executive board member of the Columbus Education Association, a senior Annenberg Civic Education Initiative Fellow, and a member of the C. J. Prentiss Emerging Leaders Project.

Álvaro Moreira Hypolito is Professor of the School of Education at the Federal University of Pelotas, Brazil. His research interests include curriculum theory, educational restructuring, and teachers' work, on which he has written articles and books. He earned his PhD in Curriculum and Instruction at the University of Wisconsin–Madison.

Larry Kuehn is Director of Research and technology for the British Columbia Teachers' Federation, of which he is a past president.

Jon Lewis is the Research and Media Officer of the South African Democratic Teachers Union (SADTU).

Shermain Mannah is the Education Officer of the South African Democratic Teachers Union (SADTU).

Ian Murch is the National Treasurer of the National Union of Teachers (NUT) in the UK. He is also the Secretary of the Bradford Division of the NUT and a leading campaigner against the creeping privatization of education in England.

Kathleen A. Murphey is Professor and Associate Dean in the School of Education, Indiana University–Purdue University Fort Wayne, Fort Wayne, Indiana.

Thulas Nxesi is the General Secretary of the South African Democratic Teachers Union (SADTU) and the President of Education International (EI).

John Nyambe is Chief Education Officer (CEO), Professional, Research and Resource Development, of the National Institute for Educational Development (NIED) in Namibia.

Nurit Peled-Elhanan is a lecturer of Language Education at the University of Tel-Aviv and the David Yellin Teachers College. Elhanan's daughter, Smadar Elhanan, was the victim of a suicide bombing in 1997. A tireless campaigner against the occupation of Palestine by Israel, she is a laureate of the 2001 Sakharov prize for Human Rights and the Freedom of Speech.

Rodolfo Rincones is an associate professor at the University of Texas at El Paso and professor at the Universidad Autónoma de Ciudad Juárez. His research interests are educational reform, policy and politics of education, educational leadership, comparative and international education, and more recently, privatization of education in México.

Susan Robertson is a Professor of Sociology of Education in the Graduate School of Education, University of Bristol. Her earlier work focused on teachers' work, state restructuring, and education policy in Australia, Canada, and New Zealand. She is author of *A Class Act: Changing Teachers' Work, the State and Globalisation* (2000, Falmer).

Harold Samuel is the chairperson of the board of trustees of the South African Democratic Teachers Union (SADTU). He was the first national treasurer of SADTU in 1990. He was involved in the struggle for the unity of educators under apartheid as president of the Teachers Association of South Africa, an Indian collective of educators.

César Silva-Montes has a doctorate in Social Sciences. He is a professor at the Universidad Autónoma de Ciudad Juárez and academic advisor at Preparatoria Altavista. His research interests are educational politics in the context of neoliberalism and globalization, specifically the changes in curriculum, evaluation, and pedagogy at public universities.

Jette Steensen is Senior Lecturer, VIA University College, Department of Teacher Education, Denmark.

Christine Stewart is also known as "Galxa'guii biik sook" a name selected by her family, which translates to "Robin Flying Through." She is a Vancouver teacher currently working for the British Columbia Teachers' Federation (BCTF), representing the union on matters concerning Aboriginal education.

She has academic training in psychology and women's studies, and she writes about the needs of indigenous peoples.

Kyle Westbrook teaches in the Chicago Public Schools and is a doctoral candidate at the University of Illinois, Chicago.

Kenneth Zeichner is Hoefs-Bascom Professor of Teacher Education at the University of Wisconsin–Madison.

Index